Advanced Animal Genetics

Advanced Animal Genetics

Ajay Chauhan

Oxford Book Company
Jaipur ▪ New Delhi

Published By

Oxford Book Company

Regd. Off. "Bony Residency", Gate No. 2
Opp. Tilak Public School, Vishweswariya Nagar,
Gopalpura Road, Jaipur - 302018, India
Ph: 0141 - 2761280, Telefax: 2761381
email - oxfordbookjpr@gmail.com
website - oxfordbookcompany.com

Delhi Office: 102, 1st Floor, Satyam House,
4327/3, Ansari Road, Darya Ganj,
New Delhi - 110002, India
Ph: 011-45652440

Distributed By
KEYA INTERNATIONAL PRESS LLP
E-mail: keyainternationalpressllp@gmail.com

First published in India, 2026

ISBN : 978-93-5524-578-6

Typeset by Shivangi Computers, Jaipur

Printed and bound in India by Thomson Press Ltd.,
New Delhi

Contents

Preface

Genetics, the cornerstone of biological inheritance, has long captivated scientists and researchers striving to decode the intricate blueprint of life. The realm of animal genetics, in particular, has emerged as a pivotal field, bridging fundamental genetic principles with their practical applications in breeding, conservation, and biotechnology. As we navigate an era defined by rapid scientific advancements and the integration of cutting-edge technologies, a deeper understanding of genetic mechanisms in animals has become indispensable.

Advanced Animal Genetics is an endeavor to present a comprehensive and refined exposition of the principles, methodologies, and applications that shape this dynamic discipline. This book delves into classical genetic theories, molecular genetics, genomic selection, and modern biotechnological interventions that have revolutionized animal breeding and conservation. By amalgamating theoretical concepts with empirical research, this volume aspires to provide a holistic perspective on animal genetics, catering to students, researchers, and professionals alike.

The journey of conceptualizing and crafting this book stemmed from a profound realization—the need for an advanced yet accessible resource that integrates foundational knowledge with contemporary advancements. Genetic engineering, CRISPR-mediated genome editing, epigenetics, and quantitative trait loci (QTL) mapping are no longer confined to theoretical discussions; they are actively transforming the landscape of livestock improvement, disease resistance, and species conservation. Hence, this book meticulously unpacks these innovations, providing readers with both conceptual clarity and practical insights.

Moreover, the ethical, ecological, and socioeconomic implications of genetic interventions in animals necessitate a well-rounded discourse. The chapters herein not only elucidate genetic mechanisms but also explore the ethical considerations, sustainability challenges, and policy frameworks that govern genetic applications. From selective breeding in livestock to genomic conservation of endangered species, the content is structured to foster a nuanced understanding of genetics in its broadest context.

This book is designed to cater to a diverse readership. For students, it offers a structured introduction to advanced genetic principles. For researchers, it presents in-depth analyses of contemporary studies and methodologies. For professionals in animal breeding, veterinary sciences, and wildlife conservation, it provides valuable insights into genetic innovations that are shaping the future.

— ***Ajay Chauhan***

Chapter 1

Molecular Mechanisms in Animals

The molecular basis of dominance was unknown to Mendel. It is now understood that a gene locus includes a long series (hundreds to thousands) of bases or nucleotides of deoxyribonucleic acid (DNA) at a particular point on a chromosome. The central dogma of molecular biology states that "DNA makes RNA makes protein", that is, that DNA is transcribed to make an RNA copy, and RNA is translated to make a protein. In this process, different alleles at a locus may or may not be transcribed, and if transcribed may be translated to slightly different forms of the same protein (called isoforms). Proteins often function as enzymes that catalyze chemical reactions in the cell, which directly or indirectly produce phenotypes. In any diploid organism, the DNA sequences of the two alleles present at any gene locus may be identical (homozygous) or different (heterozygous). Even if the gene locus is heterozygous at the level of the DNA sequence, the proteins made by each allele may be identical. In the absence of any difference between the protein products, neither allele can be said to be dominant (see co-dominance, above).

Even if the two protein products are slightly different (allozymes), it is likely that they produce the same phenotype with respect to enzyme action, and again neither allele can be said to be dominant. Dominance typically occurs when one of the two alleles is non-functional at the molecular level, that is, it is not transcribed or else does not produce a protein product. This can be the result of a mutation that alters the DNA sequence of the allele. An organism homozygous for the non-functional allele will generally show a distinctive phenotype, due to the absence of the protein product. For example, in humans and other organisms, the unpigmented skin of the albino phenotype results when an individual is homozygous for an allele that prevents synthesis of the skin pigment protein melanin. It is important to understand that it is not the lack of function that allows the allele to be described as recessive: this is the interaction with the alternative allele in the heterozygote. Three general types of interaction are possible:

1. In the typical case, the single functional allele makes sufficient protein to produce a phenotype identical to that of the homozygote: this is called haplosufficiency. For example, suppose the standard amount of enzyme produced in the functional homozygote is 100%, with the two functional alleles contributing 50% each. The single functional allele in the heterozygote produces 50% of the standard amount of enzyme, which is sufficient to produce the standard phenotype. If the heterozygote and the functional-allele homozygote have identical phenotypes, the functional allele is dominant to the non-functional allele. This occurs at the albino gene locus: the heterozygote produces sufficient enzyme to convert the pigment precursor to melanin, and the individual has standard pigmentation.
2. Alternatively, a single functional allele in the heterozygote may produce insufficient gene product for proper function, and the phenotype resembles that of the homozygote for the non-functional allele. This haploinsufficiency is much less common: usually the deficiency of gene product results in incomplete dominance.
3. The intermediate interaction occurs where the heterozygous genotype produces a phenotype intermediate between the two homozygotes. Depending on which of the two homozygotes the heterozygote most resembles, one allele is said to show incomplete dominance over the other. For example, in humans the Hb gene locus is responsible for the Beta-chain protein (HBB) that is one of the two globin proteins that make up the blood pigment hemoglobin Many people are homozygous for an allele called Hb ; some persons carry an alternative allele called Hb , either as homozygotes or heterozygotes. The hemoglobin molecules of Hb /Hb

homozygotes undergo a change in shape that distorts the morphology of the red blood cells, and causes a severe, life-threatening form of anemia called sickle-cell anemia. Persons heterozygous Hb /Hb for this allele have a much less severe form of anemia called sickle-cell trait. Because the disease phenotype of Hb /Hb heterozygotes is more similar to but not identical to the Hb /Hb homozygote, the Hb allele is said to be incompletely dominant to the Hb allele.

In some cases, dominance of a non-standard allele results when that allele produces a defective protein that interferes with the proper function of the protein produced by the standard allele. The presence of the defective protein "dominates" the standard protein, and the disease phenotype of the heterozygote more closely resembles that of the homozygote for two variant alleles. This phenomenon occurs in a number of trinucleotide repeat diseases: for an example and more details see Huntington Disease.

Dominant and Recessive Genetic Diseases in Humans

In humans, many genetic traits or diseases are classified simply as "dominant" or "recessive." Especially with respect to so-called recessive diseases, this can oversimplify the underlying molecular basis and lead to misunderstanding of the nature of dominance. For example, the genetic disease phenylketonuria (PKU) results from any of a large number (>60) of alleles at the gene locus for the enzyme phenylalanine hydroxylase (PAH). Many of these alleles produce little or no PAH, as a result of which the substrate phenylalanine and its metabolic byproducts accumulate in the central nervous system and can cause severe mental retardation if untreated. The genotypes and phenotypic consequences of interactions among three alleles are shown in the following table:

Genotype	PAH activity	[phe] conc	PKU ?
AA	*100%*	*60 uM*	*No*
AB	*30%*	*120 uM*	*No*
CC	*5%*	*200 ~ 300 uM*	
	Hyperphenylalanemia		
BB	*0.3%*	*600 ~ 2400 uM*	*Yes*

In unaffected persons homozygous for a standard functional allele (AA), PAH activity is standard (100%), and the concentration of phenylalanine in the blood [phe] is about 60 uM. In untreated persons homozygous for one of the PKU alleles (BB), PAH activity is close to zero, [phe] ten to forty times standard, and the individual manifests PKU. In the AB heterozygote, PAH activity is only 30% (not 50%) of standard, blood [phe] is elevated two-fold, and the person does not manifest PKU. Thus, the A allele is dominant to the B allele with respect to PKU, but the B allele is incompletely dominant to the A allele with respect to its molecular effect, determination of PAH activity level (0.3% < 30% << 100%). Finally, the A allele is an incomplete dominant to B with respect to [phe], as 60 uM < 120 uM << 600 uM. Note once more that it is irrelevant to the question of dominance that the recessive allele produces a more extreme [phe] phenotype. For a third allele C, a CC homozygote produces a very small amount of PAH enzyme, which results in a somewhat elevated level of [phe] in the blood, a condition called hyperphenylalanemia, which does not result in mental retardation. That is, the dominance relationships of any two alleles may vary according to which aspect of the phenotype is under consideration. It is typically more useful to talk about the phenotypic consequences of the allelic interactions involved in any genotype, rather than to try to force them into dominant and recessive categories.

History

The concept of dominance was first described by the "Father of Genetics," Gregor Mendel, in the 1860s. Mendel observed that, for a variety of traits of garden peas having to do with the appearance of seeds, seed pods, and plant appearance, there occurred two discrete phenotypes: round vs wrinkled, or yellow vs green seeds, red vs white flowers, tall vs short plants, and so on. When bred separately, the plants always produced the same phenotypes, generation after generation.

However, when lines with different phenotypes were crossed (interbred), one and only one of the parental phenotypes showed up in the offspring: green, or round, or red, or tall, and so on. However, when these hybrid plants were crossed, the offspring plants showed the two original phenotypes, in a characteristic 3:1 ratio, with the more common type having the phenotype of the parental hybrid plants. Mendel reasoned that each of the parents in the first cross were homozygotes for different alleles (AA and aa), that each contributed one allele to the offspring, such that all of these hybrids were heterozygotes (Aa), and that one of the two alleles in the hybrid cross dominated expression of the other: A masked a. The final cross between two heterozygotes (Aa X Aa) would produce AA, Aa, and aa offspring in a 1:2:1 genotype ratio with the first three classes showing the "A" phenotype, and the last showing the "a" phenotype, thereby producing the 3:1 phenotype ratio.

Mendel did not use the terms gene, allele, phenotype, genotype, homozygote, and heterozygote, all of which were introduced afterward. He did introduce the notation of capital and lowercase letters for dominant and recessive alleles, respectively, still in use today.

Mitochondrial DNA

Mitochondrial DNA (mtDNA) is the DNA located in organelles called mitochondria, structures within eukaryotic cells that convert the chemical energy from food into a form that cells can use, adenosine triphosphate (ATP).

Most other DNA present in eukaryotic organisms is found in the cell nucleus. Mitochondrial DNA can be regarded as the smallest chromosome, and was the first significant part of the human genome to be sequenced. In most species, including humans, mtDNA is inherited solely from the mother.

The DNA sequence of mtDNA has been determined from a large number of organisms and individuals (including some organisms that are extinct), and the comparison of those DNA sequences represents a mainstay of phylogenetics, in that it allows biologists to elucidate the evolutionary relationships among species. It also permits an examination of the relatedness of populations, and so has become important in anthropology and field biology.

Replication

mtDNA is replicated by the DNA polymerase gamma complex which is composed of a 140 kDa catalytic DNA polymerase encoded by the POLG gene and a 55 kDa accessory subunit encoded by the POLG2 gene. During embryogenesis, replication of mtDNA is strictly down-regulated from the fertilized oocyte through the preimplantation embryo. At the blastocyst stage, the onset of mtDNA replication is specific to the cells of the trophectoderm. In contrast, the cells of the inner cell mass restrict mtDNA replication until they receive the signals to differentiate to specific cell types.

Origin

Nuclear and mitochondrial DNA are thought to be of separate evolutionary origin, with the mtDNA being derived from the circular genomes of the bacteria that were engulfed by the early ancestors of today's eukaryotic cells. This theory is called the endosymbiotic theory. Each mitochondrion is estimated to contain 2-10 mtDNA copies. In the cells of extant organisms, the vast majority of the proteins present in the mitochondria (numbering approximately 1500 different types in mammals) are coded for by nuclear DNA, but the genes for some of them, if not most, are thought to have originally been of bacterial origin, having since been transferred to the eukaryotic nucleus during evolution.

Mitochondrial Inheritance

In most multicellular organisms, mtDNA is inherited from the mother (maternally inherited). Mechanisms for this include simple dilution (an egg contains 100,000 to 1,000,000 mtDNA molecules, whereas a sperm contains only 100 to 1000), degradation of sperm mtDNA in the fertilized egg, and, at least in a few organisms, failure of sperm mtDNA to enter the egg. Whatever the mechanism, this single parent (uniparental) pattern of mtDNA inheritance is found in most animals, most plants and in fungi as well.

Female Inheritance

In sexual reproduction, mitochondria are normally inherited exclusively from the mother. The mitochondria in mammalian sperm are usually destroyed by the egg cell after fertilization. Also, most mitochondria are present at the base of the sperm's tail, which is used for propelling the sperm cells. Sometimes the tail is lost during fertilization. In 1999 it was reported that paternal sperm mitochondria (containing mtDNA) are marked with ubiquitin to select them for later destruction inside the embryo. Some in vitro fertilization techniques, particularly injecting a sperm into an oocyte, may interfere with this. The fact that mitochondrial DNA is maternally inherited enables researchers to trace maternal lineage far back in time. (Y-chromosomal DNA, paternally inherited, is used in an analogous way to trace the agnate lineage.)

This is accomplished in humans by sequencing one or more of the hypervariable control regions (HVR1 or HVR2) of the mitochondrial DNA, as with a genealogical DNA test. HVR1 consists of about 440 base pairs. These 440 base pairs are then compared to the control regions of other individuals (either specific people or subjects in a database) to determine maternal lineage. Most often, the comparison is made to the revised Cambridge Reference Sequence. Vilà et al. have published studies tracing the matrilineal descent of domestic dogs to wolves. The concept of the Mitochondrial Eve is based on the same type of analysis, attempting to discover the origin of humanity by tracking the lineage back in time.

Because mtDNA is not highly conserved and has a rapid mutation rate, it is useful for studying the evolutionary relationships - phylogeny - of organisms. Biologists can determine and then compare mtDNA sequences among different species and use the comparisons to build an evolutionary tree for the species examined. Because mtDNA is transmitted from mother to child (both male and female), it can be a useful tool in genealogical research into a person's maternal line.

Male Inheritance

It has been reported that mitochondria can occasionally be inherited from the father in some species such as mussels. Paternally inherited mitochondria have additionally been reported in some insects such as fruit flies, honeybees, and periodical cicadas. Evidence supports rare instances of male mitochondrial inheritance in some mammals as well. Specifically, documented occurrences exist for mice, where the male-inherited mitochondria was subsequently rejected. It has also been found in sheep, and in cloned cattle. It has been found in a single case in a human male. While many of these cases involve cloned embryos or subsequent rejection of the paternal mitochondria, others document in vivo inheritance and persistence under lab conditions.

Structure

In humans (and probably in metazoans in general), 100-10,000 separate copies of mtDNA are usually present per cell (egg and sperm cells are exceptions). In mammals, each double-stranded circular mtDNA molecule consists of 15,000-17,000 base pairs. The two strands of mtDNA are differentiated by their nucleotide content with the guanine rich strand referred to as the heavy strand, and the cytosine rich strand referred to as the light strand. The heavy strand encodes 28 genes, and the light strand encodes 9 genes for a total of 37 genes. Of the 37 genes, 13 are for proteins (polypeptides), 22 are for transfer RNA (tRNA) and two are for the small and large subunits of ribosomal RNA (rRNA). This pattern is also seen among most metazoans, although in some cases one or more of the 37 genes is absent and the mtDNA size range is greater. Even greater variation in mtDNA gene content and size exists among fungi and plants, although there appears to be a core subset of genes that are present in all eukaryotes (except for the few that have no mitochondria at all).

Some plant species have enormous mtDNAs (as many as 2,500,000 base pairs per mtDNA molecule) but, surprisingly, even those huge mtDNAs contain the same number and kinds of genes as related plants with much smaller mtDNAs. The genome of the mitochondrion of the cucumber (Cucumis sativus) is three entirely or largely autonomous circular chromosomes (lengths 1556, 84 and 45 kilobases).

Use in Identification

In humans, mitochondrial DNA spans 16,569 DNA building blocks (base pairs), representing a fraction of the total DNA in cells. Unlike nuclear DNA, which is inherited from both parents and in which genes are rearranged in the process of recombination, there is usually no change in mtDNA from parent to offspring. Although mtDNA also recombines, it does so with copies of itself within the same mitochondrion. Because of this and because the mutation rate of animal mtDNA is higher than that of nuclear DNA, mtDNA is a powerful tool for tracking ancestry through females (matrilineage) and has been used in this role to track the ancestry of many species back hundreds of generations.

Human mtDNA can also be used to help identify individuals. Forensic laboratories occasionally use mtDNA comparison to identify human remains, and especially to identify older unidentified skeletal remains. Although unlike nuclear DNA, mtDNA is not specific to one individual, it can be used in combination with other evidence (anthropological evidence, circumstantial evidence, and the like) to establish identification. mtDNA is also used to exclude possible matches between missing persons and unidentified remains. Many researchers believe that mtDNA is better suited to identification of older skeletal remains than nuclear DNA because the greater number of copies of mtDNA per cell increases the chance of obtaining a useful sample, and because a match with a living relative is possible even if numerous maternal generations separate the two. American outlaw Jesse James's remains were identified using a comparison between mtDNA extracted from his remains and the mtDNA of the son of the female-line great-granddaughter of his sister. Similarly, the remains of Alexandra Feodorovna (Alix of Hesse), last Empress of Russia, and her children were identified by comparison of their mitochondrial DNA with that of Prince Philip, Duke of Edinburgh, whose maternal grandmother was Alexandra's sister Victoria of Hesse.

Similarly to identify Emperor Nicholas II remains his mitochondrial DNA was compared with that of James Carnegie, 3rd Duke of Fife, whose maternal great-grandmother Alexandra of Denmark (Queen Alexandra) was sister of Nicholas II mother Dagmar of Denmark (Empress Maria Feodorovna). The low effective population size and rapid mutation rate (in animals) makes mtDNA useful for assessing genetic relationships of individuals or groups within a species and also for identifying and quantifying the phylogeny (evolutionary relationships; phylogenetics) among different species, provided they are not too distantly related. To do this, biologists determine and then compare the mtDNA sequences from different individuals or species.

Data from the comparisons is used to construct a network of relationships among the sequences, which provides an estimate of the relationships among the individuals or species from which the mtDNAs were taken. This approach has limits that are imposed by the rate of mtDNA sequence change.

In animals, the high mutation rate makes mtDNA most useful for comparisons of individuals within species and for comparisons of species that are closely or moderately-closely related, among which the number of sequence differences can be easily counted. As the species become more distantly related, the number of sequence differences becomes very large; changes begin to accumulate on changes until an accurate count becomes impossible.

History

Mitochondrial DNA was discovered in the 1960s by Margit M. K. Nass and Sylvan Nass by electron microscopy as DNase-sensitive thread inside mitochondria, and by Ellen Haslbrunner, Hans Tuppy and Gottfried Schatz by biochemical assays on highly purified mitochondrial fractions.

Sex Linkage

Sex linkage is the phenotypic expression of an allele related to the chromosomal sex of the individual. This mode of inheritance is in contrast to the inheritance of traits on autosomal chromosomes, where both sexes have the same probability of inheritance. Since humans have many more genes on the X than the Y, there are many more X-linked traits than Y-linked traits. In mammals, the female is the homozygous sex, with two X chromosomes (XX), while the male is heterozygous, with one X and one Y chromosome (XY). Genes on the X or Y chromosome are called sex linked genes. In birds, the opposite is true: the male is the homozygous sex, having two Z chromosomes (ZZ), and the female (hen) is heterozygous, having one Z and one W chromosome (ZW).

Linkage Overview

X-linked recessive traits are expressed in all heterogametics, but are only expressed in those homogametics that are homozygous for the recessive allele. For example, an X-linked recessive allele in humans causes hemophilia. Hemophilia is much more common in males than females because males are hemizygous - they only have one copy of the gene in question - and therefore express the trait when they inherit one mutant allele. In contrast, a female must inherit two mutant alleles, a less frequent event since the mutant allele is rare in the population. Tsarevich Alexei of Russia was the most famous sufferer of X-linked hemophilia, and his disease may have played an important role in the overthrow of the imperial regime, which changed the course of history for millions of people. The incidence of recessive X-linked phenotypes in females is the square of that in males (squaring a proportion less than one gives an outcome closer to 0 than the original). If 1 in 20 males in a human population are green colour blind, then 1 in 400 females in the population are expected to be colour blind $(/_{20})*(/_{20})$.

X-linked traits are maternally inherited from carrier mothers or from an affected father. Each son born to a carrier mother has a 50% probability of inheriting the X-chromosome carrying the mutant allele. There are a few Y-linked traits; these are inherited from the father. In classical genetics, a reciprocal cross is performed to test if a trait is sex-linked.

Illustration of some X-linked heredity outcomes (A) the affected father has one X-linked dominant allele, the mother is homozygous for the recessive allele: only daughters (all) will be affected. (B) the affected mother is hemizygous with one copy of the X-linked dominant allele: both daughters and sons will have 50% probability to be affected. (C) the hemizygous mother is called "carrier" because she has one copy of the recessive allele: sons will have 50% probability to be affected, 50% of unaffected daughters will become carriers like their mother.

Autosomal Recessive Inheritance

Characteristics of Autosomal Recessive Inheritance:

• An individual will be a “carrier” if they posses one mutated allele and one normal gene copy.

• There is a 50% chance that a carrier will transmit a mutated gene to a child.

• If two carrier parents have a child there is a:

  o 25% chance that both will transmit the mutated gene; in this case, the child will inherit only mutated copies of the gene from both the mother and the father and thus will be affected with the disorder.

- o 50% chance that one carrier parent will transmit the mutated gene and the other will transmit the normal gene; in this case, the child will have one mutated gene and one normal gene and will be a carrier of the disorder.
- o 25% chance that both carrier parents will transmit the normal gene; in this case the child will have only normal genes and will not be affected and will not be a carrier.

Examples:

- cystic fibrosis,
- sickle-cell disease,
- Tay-Sachs disease,
- Niemann-Pick disease,
- spinal muscular atrophy,
- Roberts's syndrome.

Sex-linked Traits in Other aAnimals

- Calico or tortoiseshell cats, which have orange-and-black fur
- White eyes in Drosophila melanogaster flies—the first sex-linked gene ever discovered
- Evolution of dominance

The evolution of dominance concerns the evolution of genetic dominance. The central argument, that modifier genes act upon other genes to make them dominant or recessive, and that these are then themselves subject to natural selection was first proposed by the British population geneticist Ronald Fisher in 1928, and expanded upon in his book The Genetical Theory of Natural Selection.

However, Sewall Wright and J.B.S. Haldane believed that the main explanation for dominance should be based on physiological factors, and that selection for modifiers was not a primary force.

This led to a major fallout between Fisher and Wright. Subsequent works, particularly in molecular biology and biochemistry, have tended to favour Wright's view without completely excluding that the Fisherian argument may, in some circumstances, apply.

List of Mendelian Traits in Humans

In Mendelian inheritance, a child receiving a dominant allele from either parent will have the dominant form of the trait. Only those that received the recessive allele from both parents present with the recessive phenotype. Purely Mendelian traits are a tiny minority of all traits, since most phenotypic traits exhibit incomplete dominance, codominance, and contributions from many genes. The recessive phenotype may theoretically skip any number of generations, lying dormant in heterozygous "carrier" individuals until they have children with someone who also has the recessive allele and both pass it on to their child.

Examples

These traits include:

- Ability to taste phenylthiocarbamide (dominant)
- Ability to smell (bitter almond-like) hydrogen cyanide
- ACHOO syndrome (dominant)
- Albinism (recessive)
- Brachydactyly (Shortness of fingers and toes)
- Wet (dominant) or dry (recessive) earwax
- Face freckles (dominant)
- Huntingtons Disease (dominant)
- Blood type

Traits Previously Believed to be Mendelian

Some traits were previously believed to be Mendelian, but their inheritance is (probably) based on more complex genetic models, possibly involving more than one gene. These include :

- Eye color
- Hair color
- Morton's toe
- Tongue rolling
- Widow's peak (allele)
- Hitchhiker's thumb (recessive)
- Cleft chin (dominant)
- Dettached (dominant) or attached (recessive) earlobes

Autosomal Recessive

Two copies of the gene must be mutated for a person to be affected by an autosomal recessive disorder. An affected person usually has unaffected parents who each carry a single copy of the mutated gene (and are referred to as carriers). Two unaffected people who each carry one copy of the mutated gene have a 25% chance with each pregnancy of having a child affected by the disorder. Examples of this type of disorder are cystic fibrosis, sickle-cell disease, Tay-Sachs disease, Niemann-Pick disease, spinal muscular atrophy, and Roberts syndrome. Certain other phenotypes, such as wet versus dry earwax, are also determined in an autosomal recessive fashion.

X-linked Dominant

X-linked dominant inheritance, sometimes referred to as X-linked dominance, is a mode of genetic inheritance by which a dominant gene is carried on the X chromosome. As an inheritance pattern, it is less common than the X-linked recessive type. In medicine, X-linked dominant inheritance indicates that a gene responsible for a genetic disorder is located on the X chromosome, and only one copy of the allele is sufficient to cause the disorder when inherited from a parent who has the disorder.

X-linked dominant traits do not necessarily affect males more than females (unlike X-linked recessive traits). The exact pattern of inheritance varies, depending on whether the father or the mother has the trait of interest. All daughters of an affected father will also be affected but none of his sons will be affected (unless the mother is also affected). In addition, the mother of an affected son is also affected (but not necessarily the other way round). X-linked dominant disorders are caused by mutations in genes on the X chromosome. Only a few disorders have this inheritance pattern, with a prime example being X-linked hypophosphatemic rickets.

Males and females are both affected in these disorders, with males typically being more severely affected than females. Some X-linked dominant conditions such as Rett syndrome, incontinentia pigmenti type 2 and Aicardi syndrome are usually fatal in males either in utero or shortly after birth, and are therefore predominantly seen in females. Exceptions to this finding are extremely rare cases in which boys with Klinefelter syndrome (47,XXY) also inherit an X-linked dominant condition and exhibit symptoms more similar to those of a female in terms of disease severity.

The chance of passing on an X-linked dominant disorder differs between men and women. The sons of a man with an X-linked dominant disorder will all be unaffected (since they receive their father's Y chromosome), and his daughters will all inherit the condition. A woman with an X-linked dominant disorder has a 50% chance of having an affected fetus with each pregnancy, although it should be noted that in cases such as incontinentia pigmenti only female offspring are generally viable. In addition, although these conditions do not alter fertility per se, individuals with Rett syndrome or Aicardi syndrome rarely reproduce. Genetics As the X chromosome is one of the sex chromosomes (the other being the Y chromosome), X-linked inheritance is determined by the gender of the parent carrying a specific gene and can often seem complex. This is due to the fact that, typically, females have two copies of the X-chromosome, while males have only one copy.

The difference between dominant and recessive inheritance patterns also plays a role in determining the chances of a child inheriting an X-linked disorder from their parentage. Males are normally hemizygous for the X chromosome, having only one copy. As a result, X-linked dominant disorders usually show higher expressivity in males than females.

Inheritance

X-linked dominant inheritance works differently depending upon whether the mother (left image) or father (right image) is the carrier of a gene that causes a disease or disorder. In X-linked dominant inheritance, when the mother alone is the carrier of a mutated, or defective gene associated with a disease or disorder; she herself will have the disorder. Her children will inherit the disorder as follows:

- Of her daughters and sons: 50% will have the disorder, 50% will be completely unaffected. Children of both genders have an even chance of receiving either of their mother's two X chromosomes, one of which contains the defective gene in question.

When the father alone is the carrier of a defective gene associated with a disease or disorder, he too will have the disorder. His children will inherit the disorder as follows:

- Of his daughters: 100% will have the disorder, since all of his daughters will receive one copy of his single X chromosome.
- Of his sons: none will have the disorder; sons do not receive an X chromosome from their father.

If both parents were carriers of a defective gene associated with a disease or disorder, they would both have the disorder. Their children would inherit the disorder as follows:

- Of their daughters: 100% will have the disorder, since all of the daughters will receive a copy of their father's X chromosome.
- Of the sons: 50% will have the disorder, 50% will be completely unaffected. Sons have an equal chance of receiving either of their mother's X chromosomes.

In such a case, where both parents carry and thus are affected by an X-linked dominant disorder, the chance of a daughter receiving two copies of the X chromosome with the defective gene is 50%, since daughters receive one copy of the X chromosome from both parents. Were this to occur with an X-linked dominant disorder, that daughter would likely experience a more severe form. Some X-linked dominant conditions such as Aicardi Syndrome are fatal to boys, therefore only girls with these conditions survive. Similarly, individuals with Klinefelter's Syndrome are referred to as "47,XXY Males".

X-linked Hypophosphatemia

X-linked hypophosphatemia (XLH), also called X-linked dominant hypophosphatemic rickets, X-linked vitamin d-resistant rickets or hypophosphatemic vitamin d-resistant rickets (HPDR), is an X-linked dominant form of rickets (or osteomalacia) that differs from most cases of rickets in that ingestion of vitamin D is relatively ineffective. It can cause bone deformity including short stature and genu varum (bow leggedness). It is associated with a mutation in the PHEX gene sequence (Xp.22) and subsequent inactivity of the PHEX protein. The prevalence of the disease is 1:20000. The leg deformity can be treated with Ilizarov frames and CHAOS surgery.

Cause and Genetics

XLH is associated with a mutation in the PHEX gene sequence, located on the human X chromosome at location Xp22.2-p22.1. The mutation results in altered (or missing) activity of the PHEX protein, which inactivates hormone-like substances (phosphatonins) that promote phosphate excretion. The resulting excess excretion of phosphate impairs bone mineralization. Biochemically, XLH is recognized by hypophosphatemia and inappropriately low level of calcitriol (1,25-$(OH)_2$ vitamin D_3).

The disorder is inherited in an X-linked dominant manner. This means the defective gene responsible for the disorder (PHEX) is located on the X chromosome, and only one copy of the defective gene is sufficient to cause the disorder when inherited from a parent who has the disorder. Males are normally hemizygous for the X chromosome, having only one copy. As a result, X-linked dominant disorders usually show higher expressivity in males than females.

As the X chromosome is one of the sex chromosomes (the other being the Y chromosome), X-linked inheritance is determined by the gender of the parent carrying a specific gene and can often seem complex. This is because, typically, females have two copies of the X-chromosome and males have only one copy. The difference between dominant and recessive inheritance patterns also plays a role in determining the chances of a child inheriting an X-linked disorder from their parentage.

X-linked Recessive

X-linked recessive conditions are also caused by mutations in genes on the X chromosome. Males are more frequently affected than females, and the chance of passing on the disorder differs between men and women. The sons of a man with an X-linked recessive disorder will not be affected, and his daughters will carry one copy of the mutated gene. A woman who is a carrier of an X-linked recessive disorder (X X) has a 50% chance of having sons who are affected and a 50% chance of having daughters who carry one copy of the mutated gene and are therefore carriers. X-linked recessive conditions include the serious diseases Hemophilia A, Duchenne muscular dystrophy, and Lesch-Nyhan syndrome as well as common and less serious conditions such as male pattern baldness and red-green color blindness. X-linked recessive conditions can sometimes manifest in females due to skewed X-inactivation or monosomy X (Turner syndrome).

Sex Differences In Phenotype/Genotypes and Frequency

In humans, generally "men are affected and women are carriers" for two reasons. The first is the simple statistical fact that if the X-chromosomes is a population that carry a particular X-linked mutation at a frequency of 'f' (for example, 1%) then that will be the frequency that men are likely to express the mutation (since they have only one X), while women will express it at a frequency of f (for example 1% * 1% = 0.01%) since they have two X's and hence two chances to get the normal allele. Thus, X-linked mutations tend to be rare in women.

Most Common

The most common X-linked recessive disorders are:

- Color blindness; a classic example of an X-linked trait because it is easy to explain the phenotype and it's relatively common 7% to 10% of men are red-green colorblind, making the above calculations predict 0.49% to 1% of women are carriers. Its commonness may be attributable to its not being a serious disability in most cases and an actual advantage in some situations (for example, not being distracted by some of the color in color based camouflage).It is also known as daltonism.
- Hemophilia A; another famous example because it was found in European royal families who intermarried and were famous enough that their pedigrees could be established and offered in textbooks as a "famous example" of an X-linked trait that had been documented in history books before mendelian genetics was understood.
- Duchenne muscular dystrophy; muscular dystrophy associated with mutations in the dystrophin gene, characterized by rapid progression of muscle degeneration, eventually leading to loss in ambulation, respiratory failure and death.
- Becker's muscular dystrophy; milder form of Duchenne, causes slowly progressive muscle weakness of the legs and pelvis.

- Hemophilia B; a blood clotting disorder caused by a mutation of the Factor IX gene, leading to a deficiency of Factor IX. It is rarer than haemophilia A. It's also called Christmas disease
- X-linked ichthyosis; a skin condition (ichthyosis) caused by the hereditary deficiency of the steroid sulfatase (STS) enzyme that affects 1 in 2000 to 1 in 6000 males.
- X-linked agammaglobulinemia (XLA); affects the body's ability to fight infection (origin of the name: A=no, gammaglobulin=Antibody). XLA patients do not generate mature B cells. B cells are part of the immune system and normally manufacture antibodies (also called immunoglobulins) which defends the body from infections (the humoral response). Patients with untreated XLA are prone to develop serious and even fatal infections.
- Glucose-6-phosphate dehydrogenase deficiency; may exhibit nonimmune hemolytic anemia in response to a number of causes, most commonly infection or exposure to certain medications or chemicals.

Less Common Disorders

This section does not cite any references or sources. Please help improve this section by adding citations to reliable sources. Unsourced material may be challenged and removed. Theoretically, a mutation in any of the genes in the Category:Genes on chromosome. X may cause disease, but below are some notable ones, with short description of symptoms:

- Adrenoleukodystrophy; leads to progressive brain damage, failure of the adrenal glands and eventually death.
- Alport syndrome; glomerulonephritis, endstage kidney disease, and hearing loss.
- Androgen insensitivity syndrome; variable degrees of undervirilization and/or infertility in XY persons of either gender
- Barth syndrome; metabolism distortion, delayed motor skills, stamina deficiency, hypotonia, chronic fatigue, delayed growth, cardiomyopathy, and compromised immune system.

- Centronuclear myopathy; where cell nuclei are abnormally located in skeletal muscle cells. In CNM the nuclei are located at a position in the center of the cell, instead of their normal location at the periphery.
- Charcot-Marie-Tooth disease (CMTX2-3); disorder of nerves (neuropathy) that is characterized by loss of muscle tissue and touch sensation, predominantly in the feet and legs but also in the hands and arms in the advanced stages of disease.
- Coffin-Lowry syndrome; severe mental retardation sometimes associated with abnormalities of growth, cardiac abnormalities, kyphoscoliosis as well as auditory and visual abnormalities.
- Fabry disease; A lysosomal storage disease causing anhidrosis, fatigue, angiokeratomas, burning extremity pain and ocular involvement.
- Hunter's Syndrome; potentially causing hearing loss, thickening of the heart valves leading to a decline in cardiac function, obstructive airway disease, sleep apnea, and enlargement of the liver and spleen.
- Hypohidrotic ectodermal dysplasia, presenting with hypohidrosis, hypotrichosis, hypodontia
- Kabuki syndrome; multiple congenital anomalies and mental retardation.
- Kennedy disease; muscle cramps and progressive weakness
- Lesch-Nyhan syndrome; neurologic dysfunction, cognitive and behavioral disturbances including self-mutilation, and uric acid overproduction (hyperuricemia)
- Lowe Syndrome; hydrophthalmia, cataracts, intellectual disabilities, aminoaciduria, reduced renal ammonia production and vitamin D-resistant rickets
- Menkes disease; sparse and coarse hair, growth failure, and deterioration of the nervous system
- Nasodigitoacoustic syndrome; mishaped nose, brachydactyly of the distal phalanges, sensorineural deafness
- Nonsyndromic deafness and X-linked nonsyndromic deafness; hearing loss

- Norrie disease; cataracts, leukocoria along with other developmental issues in the eye
- Occipital horn syndrome; deformations in the skeleton
- Ornithine transcarbamylase deficiency; developmental delay and mental retardation. Progressive liver damage, skin lesions, and brittle hair may also be seen
- Siderius X-linked mental retardation syndrome; cleft lip and palate with mental retardation and facial dysmorphism, caused by mutations in the histone demethylase PHF8
- Simpson-Golabi-Behmel syndrome; coarse faces with protruding jaw and tongue, widened nasal bridge, and upturned nasal tip
- Spinal muscular atrophy caused by UBE1 gene mutation; weakness due to loss of the motor neurons of the spinal cord and brainstem
- Wiskott-Aldrich syndrome; eczema, thrombocytopenia, immune deficiency, and bloody diarrhea
- X-linked Severe Combined Immunodeficiency (SCID); infections, usually causing death in the first years of life
- X-linked sideroblastic anemia; skin paleness, fatigue, dizziness and enlarged spleen and liver.

Characteristics

The effects of mitochondrial disease can be quite varied. Since the distribution of the defective mitochondrial DNA may vary from organ to organ within the body, and each mutation is modulated by other genome variants, the mutation that in one individual may cause liver disease might in another person cause a brain disorder.

The severity of the specific defect may also be great or small. Some minor defects cause only "exercise intolerance", with no serious illness or disability. Defects often affect the operation of the mitochondria and multiple tissues more severely, leading to multi-system diseases. Mitochondrial diseases as a rule are worse when the defective mitochondria are present in the muscles, cerebrum, or nerves, because these cells use more energy than most other cells in the body. Although mitochondrial diseases vary greatly in presentation from person to person, several major clinical categories of these conditions have been defined, based on the most common phenotypic features, symptoms, and signs associated with the particular mutations that tend to cause them. An outstanding question and area of research is whether ATP depletion or reactive oxygen species are in fact responsible for the observed phenotypic consequences.

Causes

Mitochondrial disorders may be caused by mutations, acquired or inherited, in mitochondrial DNA (mtDNA) or in nuclear genes that code for mitochondrial components. They may also be the result of acquired mitochondrial dysfunction due to adverse effects of drugs, infections, or other environmental causes. Mitochondrial DNA inheritance behaves differently from autosomal and sexually-linked inheritance. Nuclear DNA has two copies per cell (except for sperm and egg cells), one copy being inherited from the father and the other from the mother. Mitochondrial DNA, however, is strictly inherited from the mother and each mitochondrial organelle typically contains multiple mtDNA copies. During cell division the mitochondrial DNA copies segregate randomly between the two new mitochondria, and then those new mitochondria make more copies. If only a few of the mtDNA copies inherited from the mother are defective, mitochondrial division may cause most of the defective copies to end up in just one of the new mitochondria (for more detailed inheritance patterns, see Human mitochondrial genetics). Mitochondrial disease may become clinically apparent once the number of affected mitochondria reaches a certain level; this phenomenon is called "threshold expression".

Mitochondrial DNA mutations occur frequently, due to the lack of the error checking capability that mitochondrial DNA has. This means that mitochondrial DNA disorders may occur spontaneously and relatively often. Defects in enzymes that control mitochondrial DNA replication (all of which are encoded for by genes in the nuclear DNA) may also cause mitochondrial DNA mutations. Most mitochondrial function and biogenesis is controlled by nuclear DNA. Human mitochondrial DNA encodes only 13 proteins of the respiratory chain, while most of the estimated 1,500 proteins and components targeted to mitochondria are nuclear-encoded.

Defects in nuclear-encoded mitochondrial genes are associated with hundreds of clinical disease phenotypes including anemia, dementia, hypertension, lymphoma, retinopathy, seizures, and neurodevelopmental disorders.

Treatment

Although research is ongoing, treatment options are currently limited; vitamins are frequently prescribed, though the evidence for their effectiveness is limited. Membrane penetrating antioxidants have the most important role in improving mitochondrial dysfunction. Pyruvate has been proposed recently as a treatment option. Spindle transfer, where the nuclear DNA is transferred to another healthy egg cell leaving the defective mitochondrial DNA behind, is a potential treatment procedure that has been successfully carried out on monkeys.

Using a similar pronuclear transfer technique, researchers at Newcastle University successfully transplanted healthy DNA in human eggs from women with mitochondrial disease into the eggs of women donors who were unaffected. Human genetic engineering is already being used on a small scale to allow infertile women with genetic defects in their mitochondria to have children. Embryonic mitochondrial transplant and protofection have been proposed as a possible treatment for inherited mitochondrial disease, and allotopic expression of mitochondrial proteins as a radical treatment for mtDNA mutation load.

Statistics

About 1 in 4,000 children in the United States will develop mitochondrial disease by the age of 10 years. Up to 4,000 children per year in the US are born with a type of mitochondrial disease. Because mitochondrial disorders contain many variations and subsets, some particular mitochondrial disorders are very rare. Many diseases of aging are caused by defects in mitochondrial function. Since the mitochondria are responsible for processing oxygen and converting substances from the foods we eat into energy for essential cellular functions, if there are problems with the mitochondria, it can lead to many defects for adults. These include Type 2 diabetes, Parkinson's disease, atherosclerotic heart disease, stroke, Alzheimer's disease, and cancer. Many medicines can also injure the mitochondria.

"Inside the Cell" in Dr. Neal Barnard's Program for Reversing Diabetes, Rodale Press, 2007, pp. 22 – 27, which references the Feb 12, 2004 issue of the New England Journal of Medicine, an article by Yale University researchers. Dr. Barnard also references other studies in his explanation of how, in Type 2 diabetes, the mitochondria signaling process is interrupted by fats in body cells (intramyocellular lipids) which have not been properly treated. A study at Pennington Biomedical Research Cernter in Baton Brouge, LA (Diabetes 54, 2005 1926-33) showed that this in turn partially disables the genes that produce mitochondria.

Multifactorial and Polygenic (complex) Disorders

Genetic disorders may also be complex, multifactorial, or polygenic, meaning that they are likely associated with the effects of multiple genes in combination with lifestyle and environmental factors. Multifactorial disorders include heart disease and diabetes. Although complex disorders often cluster in families, they do not have a clear-cut pattern of inheritance.

This makes it difficult to determine a person's risk of inheriting or passing on these disorders. Complex disorders are also difficult to study and treat because the specific factors that cause most of these disorders have not yet been identified. On a pedigree, polygenic diseases do tend to "run in families", but the inheritance does not fit simple patterns as with Mendelian diseases. But this does not mean that the genes cannot eventually be located and studied. There is also a strong environmental component to many of them (e.g., blood pressure).

- asthma
- autoimmune diseases such as multiple sclerosis
- cancers
- ciliopathies
- cleft palate
- diabetes
- heart disease
- hypertension
- inflammatory bowel disease
- mental retardation
- mood disorder
- obesity

Chapter 2

Latest Genomic Revolution

Comparative genomics is the study of the relationship of genome structure and function across different biological species or strains. Comparative genomics is an attempt to take advantage of the information provided by the signatures of selection to understand the function and evolutionary processes that act on genomes. While it is still a young field, it holds great promise to yield insights into many aspects of the evolution of modern species. The sheer amount of information contained in modern genomes (3.2 gigabases in the case of humans) necessitates that the methods of comparative genomics are automated. Gene finding is an important application of comparative genomics, as is discovery of new, non-coding functional elements of the genome.

Comparative genomics exploits both similarities and differences in the proteins, RNA, and regulatory regions of different organisms to infer how selection has acted upon these elements. Those elements that are responsible for similarities between different species should be conserved through time (stabilizing selection), while those elements responsible for differences among species should be divergent (positive selection). Finally, those elements that are unimportant to the evolutionary success of the organism will be unconserved (selection is neutral). One of the important goals of the field is the identification of the mechanisms of eukaryotic genome evolution. It is however often complicated by the multiplicity of events that have taken place throughout the history of individual lineages, leaving only distorted and superimposed traces in the genome of each living organism. For this reason comparative genomics studies of small model organisms (for example yeast) are of great importance to advance our understanding of general mechanisms of evolution. Having come a long way from its initial use of finding functional proteins, comparative genomics is now concentrating on finding regulatory regions and siRNA molecules. Recently, it has been discovered that distantly related species often share long conserved stretches of DNA that do not appear to code for any protein. One such ultra-conserved region, that was stable from chicken to chimp has undergone a sudden burst of change in the human lineage, and is found to be active in the developing brain of the human embryo.

Computational approaches to genome comparison have recently become a common research topic in computer science. A public collection of case studies and demonstrations is growing, ranging from whole genome comparisons to gene expression analysis. This has increased the introduction of different ideas, including concepts from systems and control, information theory, strings analysis and data mining. It is anticipated that computational approaches will become and remain a standard topic for research and teaching, while multiple courses will begin training students to be fluent in both topics.

Modelling Biological Systems

Modelling biological systems is a significant task of systems biology and mathematical biology. Computational systems biology aims to develop and use efficient algorithms, data structures, visualization and communication tools to orchestrate the integration of large quantities of biological data with the goal of computer modelling. It involves the use of computer simulations of biological systems, like cellular subsystems (such as the networks of metabolites and enzymes which comprise metabolism, signal transduction pathways and gene regulatory networks) to both analyse and visualize the complex connections of these cellular processes. It is also directly associated with bioinformatics and computational biology. Artificial life or virtual evolution attempts to understand evolutionary processes via the computer simulation of simple (artificial) life forms. It is understood that an unexpected emergent property of a complex system is a result of the interplay of the cause-and-effect among simpler, integrated parts. Biological systems manifest many important examples of emergent properties in the complex interplay of components. Traditional study of biological systems requires reductive methods in which quantities of data are gathered by category, such as concentration over time in response to a certain stimulus. Computers are critical to analysis and modelling of these data. The goal is to create accurate real-time models of a system's response to environmental and internal stimuli, such as a model of a cancer cell in order to find weaknesses in its signalling pathways, or modelling of ion channel mutations to see effects on cardiomyocytes and in turn, the function of a beating heart.

A monograph on this topic summarizes an extensive amount of published research in this area up to 1987, including subsections in the following areas: computer modelling in biology and medicine, arterial system models, neuron models, biochemical and oscillation networks, quantum automata, quantum computers in molecular biology and genetics, cancer modelling, neural nets, genetic networks, abstract relational biology, metabolic-replication systems, category theory applications in biology and medicine, automata theory, cellular automata, tessallation models and complete self-reproduction, chaotic systems in organisms, relational biology and organismic theories. This published report also includes 390 references to peer-reviewed articles by a large number of authors.

Standards

By far the most widely accepted standard format for storing and exchanging models in the field is the Systems Biology Markup Language (SBML) The SBML.org website includes a guide to many important software packages used in computational systems biology. Other markup languages with different emphases include BioPAX and CellML.

Cellular Model

Creating a cellular model has been a particularly challenging task of systems biology and mathematical biology. It involves developing efficient algorithms, data structures, visualization and communication tools to orchestrate the integration of large quantities of biological data with the goal of computer modelling. It is also directly associated with bioinformatics, computational biology and Artificial life. It involves the use of computer simulations of the many cellular subsystems such as the networks of metabolites and enzymes which comprise metabolism, signal transduction pathways and gene regulatory networks to both analyse and visualize the complex connections of these cellular processes. The complex network of biochemical reaction/transport processes and their spatial organization make the development of a predictive model of a living cell a grand challenge for the 21st century.

The eukaryotic cell cycle is very complex and is one of the most studied topics, since its misregulation leads to cancers. It is possibly a good example of a mathematical model as it deals with simple calculus but gives valid results. Two research groups have produced several models of the cell cycle simulating several organisms. They have recently produced a generic eukaryotic cell cycle model which can represent a particular eukaryote depending on the values of the parameters, demonstrating that the idiosyncrasies of the individual cell cycles are due to different protein concentrations and affinities, while the underlying mechanisms are conserved.

By means of a system of ordinary differential equations these models show the change in time (dynamical system) of the protein inside a single typical cell; this type of model is called a deterministic process (whereas a model describing a statistical distribution of protein concentrations in a population of cells is called a stochastic process).

To obtain these equations an iterative series of steps must be done: first the several models and observations are combined to form a consensus diagram and the appropriate kinetic laws are chosen to write the differential equations, such as rate kinetics for stoichiometric reactions, Michaelis-Menten kinetics for enzyme substrate reactions and Goldbeter–Koshland kinetics for ultrasensitive transcription factors, afterwards the parameters of the equations (rate constants, enzyme efficiency coefficients and Michealis constants) must be fitted to match observations; when they cannot be fitted the kinetic equation is revised and when that is not possible the wiring diagram is modified. The parameters are fitted and validated using observations of both wild type and mutants, such as protein half-life and cell size. In order to fit the parameters the differential equations need to be studied. This can be done either by simulation or by analysis. In a simulation, given a starting vector (list of the values of the variables), the progression of the system is calculated by solving the equations at each time-frame in small increments. In analysis, the proprieties of the equations are used to investigate the Behaviour of the system depending of the values of the parameters and variables. A system of differential equations can be represented as a vector field, where each vector described the change (in concentration of two or more protein) determining where and how fast the trajectory (simulation) is heading.

Vector fields can have several special points: a stable point, called a sink, that attracts in all directions (forcing the concentrations to be at a certain value), an unstable point, either a source or a saddle point which repels (forcing the concentrations to change away from a certain value), and a limit cycle, a closed trajectory towards which several trajectories spiral towards (making the concentrations oscillate).

A better representation which can handle the large number of variables and parameters is called a bifurcation diagram (bifurcation theory): the presence of these special steady-state points at certain values of a parameter (e.g. mass) is represented by a point and once the parameter passes a certain value, a qualitative change occurs, called a bifurcation, in which the nature of the space changes, with profound consequences for the protein concentrations: the cell cycle has phases (partially corresponding to G1 and G2) in which mass, via a stable point, controls cyclin levels, and phases (S and M phases) in which the concentrations change independently, but once the phase has changed at a bifurcation event (cell cycle checkpoint), the system cannot go back to the previous levels since at the current mass the vector field is profoundly different and the mass cannot be reversed back through the bifurcation event, making a checkpoint irreversible. In particular the S and M checkpoints are regulated by means of special bifurcations called a Hopf bifurcation and an infinite period bifurcation.

Protein Structure Prediction

Protein structure prediction is the prediction of the three-dimensional structure of a protein from its amino acid sequence — that is, the prediction of its secondary, tertiary, and quaternary structure from its primary structure. Structure prediction is fundamentally different from the inverse problem of protein design. Protein structure prediction is one of the most important goals pursued by bioinformatics and theoretical chemistry; it is highly important in medicine (for example, in drug design) and biotechnology (for example, in the design of novel enzymes). Every two years, the performance of current methods is assessed in the CASP experiment (Critical Assessment of Techniques for Protein Structure Prediction).

Secondary Structure

Secondary structure prediction is a set of techniques in bioinformatics that aim to predict the local secondary structures of proteins and RNA sequences based only on knowledge of their primary structure-amino acid or nucleotide sequence, respectively. For proteins, a prediction consists of assigning regions of the amino acid sequence as likely alpha helices, beta strands (often noted as "extended" conformations), or turns. The success of a prediction is determined by comparing it to the results of the DSSP algorithm applied to the crystal structure of the protein; for nucleic acids, it may be determined from the hydrogen bonding pattern. Specialized algorithms have been developed for the detection of specific well-defined patterns such as transmembrane helices and coiled coils in proteins, or canonical microRNA structures in RNA.

The best modern methods of secondary structure prediction in proteins reach about 80% accuracy; this high accuracy allows the use of the predictions in fold recognition and ab initio protein structure prediction, classification of structural motifs, and refinement of sequence alignments. The accuracy of current protein secondary structure prediction methods is assessed in weekly benchmarks such as LiveBench and EVA.

Background

Early methods of secondary structure prediction, introduced in the 1960s and early 1970s, focused on identifying likely alpha helices and were based mainly on helix-coil transition models. Significantly more accurate predictions that included beta sheets were introduced in the 1970s and relied on statistical assessments based on probability parameters derived from known solved structures. These methods, applied to a single sequence, are typically at most about 60-65% accurate, and often underpredict beta sheets.

The evolutionary conservation of secondary structures can be exploited by simultaneously assessing many homologous sequences in a multiple sequence alignment, by calculating the net secondary structure propensity of an aligned column of amino acids. In concert with larger databases of known protein structures and modern machine learning methods such as neural nets and support vector machines, these methods can achieve up 80% overall accuracy in globular proteins.

The theoretical upper limit of accuracy is around 90%, partly due to idiosyncrasies in DSSP assignment near the ends of secondary structures, where local conformations vary under native conditions but may be forced to assume a single conformation in crystals due to packing constraints. Limitations are also imposed by secondary structure prediction's inability to account for tertiary structure; for example, a sequence predicted as a likely helix may still be able to adopt a beta-strand conformation if it is located within a beta-sheet region of the protein and its side chains pack well with their neighbours. Dramatic conformational changes related to the protein's function or environment can also alter local secondary structure.

Chou-Fasman Method

The Chou-Fasman method are an empirical technique for the prediction of secondary structures in proteins, originally developed in the 1970s. The method is based on analyses of the relative frequencies of each amino acid in alpha helices, beta sheets, and turns based on known protein structures solved with X-ray crystallography. From these frequencies a set of probability parameters were derived for the appearance of each amino acid in each secondary structure type, and these parameters are used to predict the probability that a given sequence of amino acids would form a helix, a beta strand, or a turn in a protein. The method is at most about 50-60% accurate in identifying correct secondary structures, which is significantly less accurate than the GOR method or modern machine learning-based techniques.

Amino Acid Propensities

The original Chou-Fasman parameters found some strong tendencies among individual amino acids to prefer one type of secondary structure over others. Alanine, glutamate, leucine, and methionine were identified as helix formers, while proline and glycine, due to the unique conformational properties of their peptide bonds, commonly end a helix. The original Chou-Fasman parameters were derived from a very small and non-representative sample of protein structures due to the small number of such structures that were known at the time of their original work. These original parameters have since been shown to be unreliable and have been updated from a current dataset, along with modifications to the initial algorithm.

The Chou-Fasman method takes into account only the probability that each individual amino acid will appear in a helix, strand, or turn. Unlike the more complex GOR method, it does not reflect the conditional probabilities of an amino acid to form a particular secondary structure given that its neighbours already possess that structure. This lack of cooperativity increases its computational efficiency but decreases its accuracy, since the propensities of individual amino acids are often not strong enough to render a definitive prediction.

Algorithm

The Chou-Fasman method predicts helices and strands in a similar fashion, first searching linearly through the sequence for a "nucleation" region of high helix or strand probability and then extending the region until a subsequent four-residue window carries a probability of less than 1. As originally described, four out of any six contiguous amino acids were sufficient to nucleate helix, and three out of any contiguous five were sufficient for a sheet. The probability thresholds for helix and strand nucleations are constant but not necessarily equal; originally 1.03 was set as the helix cutoff and 1.00 for the strand cutoff. Turns are also evaluated in four-residue windows, but are calculated using a multi-step procedure because many turn regions contain amino acids that could also appear in helix or sheet regions. Four-residue turns also have their own characteristic amino acids; proline and glycine are both common in turns. A turn is predicted only if the turn probability is greater than the helix or sheet probabilities *and* a probability value based on the positions of particular amino acids in the turn exceeds a predetermined threshold. The turn probability p(t) is determined as:

$$p(t) = p_t(j) \times p_t(j + 1) \times p_t(j + 2) \times p_t\ (j + 3)$$

where *j* is the position of the amino acid in the four-residue window. If p(t) exceeds an arbitrary cutoff value (originally 7.5e-3), the mean of the p(j)'s exceeds 1, and p(t) exceeds the alpha helix and beta sheet probabilities for that window, then a turn is predicted. If the first two conditions are met but the probability of a beta sheet p(b) exceeds p(t), then a sheet is predicted instead.

GOR Method

The GOR method (Garnier-Osguthorpe-Robson) is an information theory-based method for the prediction of secondary structures in proteins. It was developed in the late 1970's shortly after the simpler Chou-Fasman method. Like Chou-Fasman, the GOR method is based on probability parameters derived from empirical studies of known protein tertiary structures solved by X-ray crystallography. However, unlike Chou-Fasman, the GOR method takes into account not only the propensities of individual amino acids to form particular secondary structures, but also the conditional probability of the amino acid to form a secondary structure given that its immediate neighbours have already formed that structure. The method is therefore essentially Bayesian in its analysis.

The GOR method analyses sequences to predict alpha helix, beta sheet, turn, or random coil secondary structure at each position based on 17-amino acid sequence windows. The original description of the method included four scoring matrices of size 17x20, where the columns correspond to the log-odds score, which reflects the probability of finding a given amino acid at each position in the 17-residue sequence. The four matrices reflect the probabilities of the central, ninth amino acid being in a helical, sheet, turn, or coil conformation. In subsequent revisions to the method, the turn matrix was eliminated due to the high variability of sequences in turn regions (particularly over such a large window). The method was considered as best requiring at least four contiguous residues to score as alpha helices to classify the region as helical, and at least two contiguous residues for a beta sheet.

The mathematics and algorithm of the GOR method were based on a earlier series of studies by Robson and colleagues reported mainly in the Journal of Molecular Biology (e.g.) and The Biochemical Journal (e.g.). The latter describes the information theoretic expansions in terms of conditional information measures. The use of the word "simple" in the title of the GOR paper reflected the fact that the above earlier methods provided proofs and techniques somewhat daunting by being rather unfamilar in protein science in the early 1970s; even Bayes methods were then unfamilar and controversial.

An important feature of these early studies, which survived in the GOR method, was the treatment of the sparse protein sequence data of the early 1970s by expected information measures. That is, expectations on a Bayesian basis considering the distribution of plausible information measure values given the actual frequencies (numbers of observations). The expectation measures resulting from integration over this and similar distributions may now be seen as composed of "incomplete" or extended zeta functions, e.g. z(s,observed frequency) – z(s,expected frequency) with incomplete zeta function $z(s, n) = 1 + (1/2)^6 + (1/3)^s + (1/4)^s + ... +(1/n)^s$. The GOR method used s=1. Also, in the GOR method and the earlier methods, the measure for the contrary state to e.g. helix H, i.e. ~H, was subtracted from that for H, and similarly for beta sheet, turns, and coil or loop.

Thus the method can be seen as employing a zeta function estimate of log predictive odds. An adjustable decision constant could also be applied, which thus also implies a decision theory approach; the GOR method allowed the option to use decision constants to optimize predictions for different classes of protein. The expected information measure used as a basis for the information expansion was less important by the time of publication of the GOR method because protein sequence data became more plentiful, at least for the terms considered at that time.

Then, for s=1, the expression z(s,observed frequency) – z(s,expected frequency) approaches the natural logarithm of (observed frequency/expected frequency) as frequencies increase. However, this measure (including use of other values of s) remains important in later more general applications with high dimensional data, where data for more complex terms in the information expansion are inevitably sparse (e.g.).

Machine Learning

Neural network methods use training sets of solved structures to identify common sequence motifs associated with particular arrangements of secondary structures. These methods are over 70% accurate in their predictions, although beta strands are still often underpredicted due to the lack of three-dimensional structural information that would allow assessment of hydrogen bonding patterns that can promote formation of the extended conformation required for the presence of a complete beta sheet. Support vector machines have proven particularly useful for predicting the locations of turns, which are difficult to identify with statistical methods. The requirement of relatively small training sets has also been cited as an advantage to avoid overfitting to existing structural data.

Extensions of machine learning techniques attempt to predict more fine-grained local properties of proteins, such as backbone dihedral angles in unassigned regions. Both SVMs and neural networks have been applied to this problem.

Other Improvements

It is reported that in addition to the protein sequence, secondary structure formation depends on other factors. For example, it is reported that secondary structure tendencies depend also on local environment, solvent accessibility of residues, protein structural class, and even the organism from which the proteins are obtained. Based on such observations, some studies have shown that secondary structure prediction can be improved by addition of information about protein structural class, solvent accessibility and also contact number of residues. Sequence covariation methods rely on the existence of a data set composed of multiple homologous RNA sequences with related but dissimilar sequences. These methods analyse the covariation of individual base sites in evolution; maintenance at two widely separated sites of a pair of base-pairing nucleotides indicates the presence of a structurally required hydrogen bond between those positions. The general problem of pseudoknot prediction has been shown to be NP-complete.

Tertiary Structure

The practical role of protein structure prediction is now more important than ever. Massive amounts of protein sequence data are produced by modern large-scale DNA sequencing efforts such as the Human Genome Project. Despite community-wide efforts in structural genomics, the output of experimentally determined protein structures—typically by time-consuming and relatively expensive X-ray crystallography or NMR spectroscopy—is lagging far behind the output of protein sequences. The protein structure prediction remains an extremely difficult and unresolved undertaking. The two main problems are calculation of protein free energy and finding the global minimum of this energy. A protein structure prediction method must explore the space of possible protein structures which is astronomically large.

These problems can be partially bypassed in "comparative" or homology modelling and fold recognition methods, when the search space is pruned by the assumption that the protein in question adopts a structure that is close to the experimentally determined structure of another homologous protein. On the other hand, the *de novo* or ab initio protein structure prediction methods must explicitly resolve these problems.

De Novo Protein Structure Prediction

In computational biology, *de novo* protein structure prediction is the task of estimating a protein's tertiary structure from its sequence alone. The problem is very difficult and has occupied leading scientists for decades. Research has focused in three areas: alternate lower-resolution representations of proteins, accurate energy functions, and efficient sampling methods. At present, the most successful methods have a reasonable probability of predicting the fold of a small protein domain within 5 angstroms. *De novo* protein structure prediction methods attempt to predict tertiary structures from sequences based on general principles that govern protein folding energetics and/or statistical tendencies of conformational features that native structures acquire, without the use of explicit templates.

A general paradigm for *de novo* prediction involves sampling conformation space, guided by scoring functions and other sequence-dependent biases such that a large set of candidate ("decoy") structures are generated. Native-like conformations are then selected from these decoys using scoring functions as well as conformer clustering. High-resolution refinement is sometimes used as a final step to fine-tune native-like structures. There are two major classes of scoring functions. Physics-based functions are based on mathematical models describing aspects of the known physics of molecular interaction. Knowledge-based functions are formed with statistical models capturing aspects of the properties of native protein conformations.

De novo methods tend to require vast computational resources, and have thus only been carried out for relatively small proteins. To predict protein structure *de novo* for larger proteins will require better algorithms and larger computational resources like those afforded by either powerful supercomputers (such as Blue Gene or MDGRAPE-3) or distributed computing projects (such as Rosetta@home, the Human Proteome Folding Project, or Nutritious Rice for the World). Although computational barriers are vast, the potential benefits of structural genomics (by predicted or experimental methods) make *de novo* structure prediction an active research field.

Comparative Protein Modelling

Comparative protein modelling uses previously solved structures as starting points, or templates. This is effective because it appears that although the number of actual proteins is vast, there is a limited set of tertiary structural motifs to which most proteins belong. It has been suggested that there are only around 2,000 distinct protein folds in nature, though there are many millions of different proteins.

Homology Modelling

Homology modelling, also known as comparative modelling of protein refers to constructing an atomic-resolution model of the "*target*" protein from its amino acid sequence and an experimental three-dimensional structure of a related homologous protein (the "*template*"). Homology modelling relies on the identification of one or more known protein structures likely to resemble the structure of the query sequence, and on the production of an alignment that maps residues in the query sequence to residues in the template sequence. It has been shown that protein structures are more conserved than protein sequences amongst homologues, but sequences falling below a 20% sequence identity can have very different structure.

Evolutionarily related proteins have similar sequences and naturally occurring homologous proteins have similar protein structure. It has been shown that three-dimensional protein structure is evolutionarily more conserved than expected due to sequence conservation. The sequence alignment and template structure are then used to produce a structural model of the target. Because protein structures are more conserved than DNA sequences, detectable levels of sequence similarity usually imply significant structural similarity.

The quality of the homology model is dependent on the quality of the sequence alignment and template structure. The approach can be complicated by the presence of alignment gaps (commonly called indels) that indicate a structural region present in the target but not in the template, and by structure gaps in the template that arise from poor resolution in the experimental procedure (usually X-ray crystallography) used to solve the structure. Model quality declines with decreasing sequence identity; a typical model has ~1-2 Å root mean square deviation between the matched C atoms at 70% sequence identity but only 2-4 Å agreement at 25% sequence identity. However, the errors are significantly higher in the loop regions, where the amino acid sequences of the target and template proteins may be completely different.

Regions of the model that were constructed without a template, usually by loop modelling, are generally much less accurate than the rest of the model. Errors in side chain packing and position also increase with decreasing identity, and variations in these packing configurations have been suggested as a major reason for poor model quality at low identity. Taken together, these various atomic-position errors are significant and impede the use of homology models for purposes that require atomic-resolution data, such as drug design and protein-protein interaction predictions; even the quaternary structure of a protein may be difficult to predict from homology models of its subunit(s).

Nevertheless, homology models can be useful in reaching *qualitative* conclusions about the biochemistry of the query sequence, especially in formulating hypotheses about why certain residues are conserved, which may in turn lead to experiments to test those hypotheses. For example, the spatial arrangement of conserved residues may suggest whether a particular residue is conserved to stabilize the folding, to participate in binding some small molecule, or to foster association with another protein or nucleic acid.

Homology modelling can produce high-quality structural models when the target and template are closely related, which has inspired the formation of a structural genomics consortium dedicated to the production of representative experimental structures for all classes of protein folds. The chief inaccuracies in homology modelling, which worsen with lower sequence identity, derive from errors in the initial sequence alignment and from improper template selection. Like other methods of structure prediction, current practice in homology modelling is assessed in a biannual large-scale experiment known as the Critical Assessment of Techniques for Protein Structure Prediction, or CASP.

Motive

The method of homology modelling is based on the observation that protein tertiary structure is better conserved than amino acid sequence. Thus, even proteins that have diverged appreciably in sequence but still share detectable similarity will also share common structural properties, particularly the overall fold. Because it is difficult and time-consuming to obtain experimental structures from methods such as X-ray crystallography and protein NMR for every protein of interest, homology modelling can provide useful structural models for generating hypotheses about a protein's function and directing further experimental work.

There are exceptions to the general rule that proteins sharing significant sequence identity will share a fold. For example, a judiciously chosen set of mutations of less than 50% of a protein can cause the protein to adopt a completely different fold. However, such a massive structural rearrangement is unlikely to occur in evolution, especially since the protein is usually under the constraint that it must fold properly and carry out its function in the cell. Consequently, the roughly folded structure of a protein (its "topology") is conserved longer than its amino-acid sequence and much longer than the corresponding DNA sequence; in other words, two proteins may share a similar fold even if their evolutionary relationship is so distant that it cannot be discerned reliably. For comparison, the function of a protein is conserved much *less* than the protein sequence, since relatively few changes in amino-acid sequence are required to take on a related function.

Steps in Model Production

The homology modelling procedure can be broken down into four sequential steps: template selection, target-template alignment, model construction, and model assessment. The first two steps are often essentially performed together, as the most common methods of identifying templates rely on the production of sequence alignments; however, these alignments may not be of sufficient quality because database search techniques prioritize speed over alignment quality. These processes can be performed iteratively to improve the quality of the final model, although quality assessments that are not dependent on the true target structure are still under development. Optimizing the speed and accuracy of these steps for use in large-scale automated structure prediction is a key component of structural genomics initiatives, partly because the resulting volume of data will be too large to process manually and partly because the goal of structural genomics requires providing models of reasonable quality to researchers who are not themselves structure prediction experts.

Template Selection and Sequence Alignment

The critical first step in homology modelling is the identification of the best template structure, if indeed any are available. The simplest method of template identification relies on serial pairwise sequence alignments aided by database search techniques such as FASTA and BLAST. More sensitive methods based on multiple sequence alignment-of which PSI-BLAST is the most common example-iteratively update their position-specific scoring matrix to successively identify more distantly related homologs. This family of methods has been shown to produce a larger number of potential templates and to identify better templates for sequences that have only distant relationships to any solved structure. Protein threading, also known as fold recognition or 3D-1D alignment, can also be used as a search technique for identifying templates to be used in traditional homology modelling methods. When performing a

BLAST search, a reliable first approach is to identify hits with a sufficiently low E-value, which are considered sufficiently close in evolution to make a reliable homology model.

Other factors may tip the balance in marginal cases; for example, the template may have a function similar to that of the query sequence, or it may belong to a homologous operon. However, a template with a poor E-value should generally not be chosen, even if it is the only one available, since it may well have a wrong structure, leading to the production of a misguided model. A better approach is to submit the primary sequence to fold-recognition servers or, better still, consensus meta-servers which improve upon individual fold-recognition servers by identifying similarities (consensus) among independent predictions.

Often several candidate template structures are identified by these approaches. Although some methods can generate hybrid models from multiple templates, most methods rely on a single template. Therefore, choosing the best template from among the candidates is a key step, and can affect the final accuracy of the structure significantly. This choice is guided by several factors, such as the similarity of the query and template sequences, of their functions, and of the predicted query and observed template secondary structures. Perhaps most importantly, the *coverage* of the aligned regions: the fraction of the query sequence structure that can be predicted from the template, and the plausibility of the resulting model. Thus, sometimes several homology models are produced for a single query sequence, with the most likely candidate chosen only in the final step.

It is possible to use the sequence alignment generated by the database search technique as the basis for the subsequent model production; however, more sophisticated approaches have also been explored. One proposal generates an ensemble of stochastically defined pairwise alignments between the target sequence and a single identified template as a means of exploring "alignment space" in regions of sequence with low local similarity. "Profile-profile" alignments that first generate a sequence profile of the target and systematically compare it to the sequence profiles of solved structures; the coarse-graining inherent in the profile construction is thought to reduce noise introduced by sequence drift in nonessential regions of the sequence.

Model Generation

Given a template and an alignment, the information contained therein must be used to generate a three-dimensional structural model of the target, represented as a set of Cartesian coordinates for each atom in the protein. Three major classes of model generation methods have been proposed.

Fragment Assembly

The original method of homology modelling relied on the assembly of a complete model from conserved structural fragments identified in closely related solved structures. For example, a modelling study of serine proteases in mammals identified a sharp distinction between "core" structural regions conserved in all experimental structures in the class, and variable regions typically located in the loops where the majority of the sequence differences were localized. Thus unsolved proteins could be modelled by first constructing the conserved core and then substituting variable regions from other proteins in the set of solved structures. Current implementations of this method differ mainly in the way they deal with regions that are not conserved or that lack a template. The variable regions are often constructed with the help of fragment libraries.

Segment Matching

The segment-matching method divides the target into a series of short segments, each of which is matched to its own template fitted from the Protein Data Bank. Thus, sequence alignment is done over segments rather than over the entire protein. Selection of the template for each segment is based on sequence similarity, comparisons of alpha carbon coordinates, and predicted steric conflicts arising from the van der Waals radii of the divergent atoms between target and template.

Satisfaction of Spatial Restraints

The most common current homology modelling method takes its inspiration from calculations required to construct a three-dimensional structure from data generated by NMR spectroscopy. One or more target-template alignments are used to construct a set of geometrical criteria that are then converted to probability density functions for each restraint. Restraints applied to the main protein internal coordinates-protein backbone distances and dihedral angles-serve as the basis for a global optimization procedure that originally used conjugate gradient energy minimization to iteratively refine the positions of all heavy atoms in the protein. This method had been dramatically expanded to apply specifically to loop modelling, which can be extremely difficult due to the high flexibility of loops in proteins in aqueous solution. A more recent expansion applies the spatial-restraint model to electron density maps derived from cryoelectron microscopy studies, which provide low-resolution information that is not usually itself sufficient to generate atomic-resolution structural models. To address the problem of inaccuracies in initial target-template sequence alignment, an iterative procedure has also been introduced to refine the alignment on the basis of the initial structural fit. The most commonly used software in spatial restraint-based modelling is MODELLER and a database called ModBase has been established for reliable models generated with it.

Loop Modelling

Regions of the target sequence that are not aligned to a template are modelled by loop modelling; they are the most susceptible to major modelling errors and occur with higher frequency when the target and template have low sequence identity. The coordinates of unmatched sections determined by loop modelling programs are generally much less accurate than those obtained from simply copying the coordinates of a known structure, particularly if the loop is longer than 10 residues.

The first two sidechain dihedral angles ($\div_1$ and $\div_2$) can usually be estimated within 30° for an accurate backbone structure; however, the later dihedral angles found in longer side chains such as lysine and arginine are notoriously difficult to predict. Moreover, small errors in $\div_1$ (and, to a lesser extent, in $\div_2$) can cause relatively large errors in the positions of the atoms at the terminus of side chain; such atoms often have a functional importance, particularly when located near the active site.

Model Assessment

Assessment of homology models without reference to the true target structure is usually performed with two methods:

statistical potentials or physics-based energy calculations. Both methods produce an estimate of the energy (or an energy-like analog) for the model or models being assessed; independent criteria are needed to determine acceptable cutoffs. Neither of the two methods correlates exceptionally well with true structural accuracy, especially on protein types underrepresented in the PDB, such as membrane proteins.

Statistical potentials are empirical methods based on observed residue-residue contact frequencies among proteins of known structure in the PDB. They assign a probability or energy score to each possible pairwise interaction between amino acids and combine these pairwise interaction scores into a single score for the entire model. Some such methods can also produce a residue-by-residue assessment that identifies poorly scoring regions within the model, though the model may have a reasonable score overall. These methods emphasize the hydrophobic core and solvent-exposed polar amino acids often present in globular proteins. Examples of popular statistical potentials include Prosa and DOPE. Statistical potentials are more computationally efficient than energy calculations.

Physics-based energy calculations aim to capture the interatomic interactions that are physically responsible for protein stability in solution, especially van der Waals and electrostatic interactions. These calculations are performed using a molecular mechanics force field; proteins are normally too large even for semi-empirical quantum mechanics-based calculations. The use of these methods is based on the energy landscape hypothesis of protein folding, which predicts that a protein's native state is also its energy minimum. Such methods usually employ implicit solvation, which provides a continuous approximation of a solvent bath for a single protein molecule without necessitating the explicit representation of individual solvent molecules. A force field specifically constructed for model assessment is known as the Effective Force Field (EFF) and is based on atomic parameters from CHARMM.

A very extensive model validation report can be obtained using the Radboud Universiteit Nijmegen *"What Check"* software which is one option of the Radboud Universiteit Nijmegen *"What If"* software package; it produces a many page document with extensive analyses of nearly 200 scientific and administrative aspects of the model. *"What Check"* is available as a free server; it can also be used to validate experimentally determined structures of macromolecules.

One newer method for model assessment relies on machine learning techniques such as neural nets, which may be trained to assess the structure directly or to form a consensus among multiple statistical and energy-based methods. Very recent results using support vector machine regression on a jury of more traditional assessment methods outperformed common statistical, energy-based, and machine learning methods.

Structural Comparison Methods

The assessment of homology models' accuracy is straightforward when the experimental structure is known. The most common method of comparing two protein structures uses the root-mean-square deviation (RMSD) metric to measure the mean distance between the corresponding atoms in the two structures after they have been superimposed. However, RMSD does underestimate the accuracy of models in which the core is essentially correctly modelled, but some flexible loop regions are inaccurate. A method introduced for the modelling assessment experiment CASP is known as the global distance test (GDT) and measures the total number of atoms whose distance from the model to the experimental structure lies under a certain distance cutoff. Both methods can be used for any subset of atoms in the structure, but are often applied to only the alpha carbon or protein backbone atoms to minimize the noise created by poorly modelled side chain rotameric states, which most modelling methods are not optimized to predict.

Benchmarking

Several large-scale benchmarking efforts have been made to assess the relative quality of various current homology modelling methods. CASP is a community-wide prediction experiment that runs every two years during the summer months and challenges prediction teams to submit structural models for a number of sequences whose structures have recently been solved experimentally but have not yet been published. Its partner CAFASP has run in parallel with CASP but evaluates only models produced via fully automated servers. Continuously running experiments that do not have prediction 'seasons' focus mainly on benchmarking publicly available webservers. LiveBench and EVA run continuously to assess participating servers' performance in prediction of imminently released structures from the PDB. CASP and CAFASP serve mainly as evaluations of the state of the art in modelling, while the continuous assessments seek to evaluate the model quality that would be obtained by a non-expert user employing publicly available tools.

Accuracy

The accuracy of the structures generated by homology modelling is highly dependent on the sequence identity between target and template. Above 50% sequence identity, models tend to be reliable, with only minor errors in side chain packing and rotameric state, and an overall RMSD between the modelled and the experimental structure falling around 1 Â. This error is comparable to the typical resolution of a structure solved by NMR. In the 30-50% identity range, errors can be more severe and are often located in loops.

Below 30% identity, serious errors occur, sometimes resulting in the basic fold being mis-predicted. This low-identity region is often referred to as the "twilight zone" within which homology modelling is extremely difficult, and to which it is possibly less suited than fold recognition methods. At high sequence identities, the primary source of error in homology modelling derives from the choice of the template or templates on which the model is based, while lower identities exhibit serious errors in sequence alignment that inhibit the production of high-quality models. It has been suggested that the major impediment to quality model production is inadequacies in sequence alignment, since "optimal" structural alignments between two proteins of known structure can be used as input to current modelling methods to produce quite accurate reproductions of the original experimental structure.

Attempts have been made to improve the accuracy of homology models built with existing methods by subjecting them to molecular dynamics simulation in an effort to improve their RMSD to the experimental structure. However, current force field parameterizations may not be sufficiently accurate for this task, since homology models used as starting structures for molecular dynamics tend to produce slightly worse structures. Slight improvements have been observed in cases where significant restraints were used during the simulation.

Sources of Error

The two most common and large-scale sources of error in homology modelling are poor template selection and inaccuracies in target-template sequence alignment. Controlling for these two factors by using a structural alignment, or a sequence alignment produced on the basis of comparing two solved structures, dramatically reduces the errors in final models; these "gold standard" alignments can be used as input to current modelling methods to produce quite accurate reproductions of the original experimental structure. Results from the most recent CASP experiment suggest that "consensus" methods collecting the results of multiple fold recognition and multiple alignment searches increase the likelihood of identifying the correct template; similarly, the use of multiple templates in the model-building step may be less optimal than the use of the single correct template but more optimal than the use of a single suboptimal one. Alignment errors may be minimized by the use of a multiple alignment even if only one template is used, and by the iterative refinement of local regions of low similarity. A lesser source of model errors are errors in the template structure. The PDBREPORT database lists several million, mostly very small but occasionally dramatic, errors in experimental (template) structures that have been deposited in the PDB. Serious local errors can arise in homology models where an insertion or deletion mutation or a gap in a solved structure result in a region of target sequence for which there is no corresponding template.

This problem can be minimized by the use of multiple templates, but the method is complicated by the templates' differing local structures around the gap and by the likelihood that a missing region in one experimental structure is also missing in other structures of the same protein family. Missing regions are most common in loops where high local flexibility increases the difficulty of resolving the region by structure-determination methods. Although some guidance is provided even with a single template by the positioning of the ends of the missing region, the longer the gap, the more difficult it is to model. Loops of up to about 9 residues can be modelled with moderate accuracy in some cases if the local alignment is correct. Larger regions are often modelled individually using ab initio structure prediction techniques, although this approach has met with only isolated success.

The rotameric states of side chains and their internal packing arrangement also present difficulties in homology modelling, even in targets for which the backbone structure is relatively easy to predict. This is partly due to the fact that many side chains in crystal structures are not in their "optimal" rotameric state as a result of energetic factors in the hydrophobic core and in the packing of the individual molecules in a protein crystal. One method of addressing this problem requires searching a rotameric library to identify locally low-energy combinations of packing states. It has been suggested that a major reason that homology modelling so difficult when target-template sequence identity lies below 30% is that such proteins have broadly similar folds but widely divergent side chain packing arrangements.

Utility

Uses of the structural models include protein-protein interaction prediction, protein-protein docking, molecular docking, and functional annotation of genes identified in an organism's genome. Even low-accuracy homology models can be useful for these purposes, because their inaccuracies tend to be located in the loops on the protein surface, which are normally more variable even between closely related proteins. The functional regions of the protein, especially its active site, tend to be more highly conserved and thus more accurately modelled. Homology models can also be used to identify subtle differences between related proteins that have not all been solved structurally.

For example, the method was used to identify cation binding sites on the Na^+/K^+ ATPase and to propose hypotheses about different ATPases' binding affinity. Used in conjunction with molecular dynamics simulations, homology models can also generate hypotheses about the kinetics and dynamics of a protein, as in studies of the ion selectivity of a potassium channel. Large-scale automated modelling of all identified protein-coding regions in a genome has been attempted for the yeast *Saccharomyces cerevisiae*, resulting in nearly 1000 quality models for proteins whose structures had not yet been determined at the time of the study, and identifying novel relationships between 236 yeast proteins and other previously solved structures.

Threading (Protein Sequence)

Protein threading, also known as fold recognition, is a method of protein modelling (i.e. computational protein structure prediction) which is used to model those proteins which have the same fold as proteins of known structures, but do not have homologous proteins with known structure. It differs from the homology modelling method of structure prediction in that it is used for proteins which do not have their homologous protein structures deposited in the Protein Data Bank (PDB), whereas homology modelling is used for those proteins which do. Threading works by using statistical knowledge of the relationship between the structures deposited in the PDB and the sequence of the protein which one wishes to model.

The prediction is made by "threading" (i.e. placing, aligning) each amino acid in the target sequence to a position in the template structure, and evaluating how well the target fits the template. After the best-fit template is selected, the structural model of the sequence is built based on the alignment with the chosen template. Protein threading is based on two basic observations: that the number of different folds in nature is fairly small (approximately 1000); and that 90% of the new structures submitted to the PDB in the past three years have similar structural folds to ones already in the PDB.

Classification of Protein Structure

The Structural Classification of Proteins (SCOP) database provides a detailed and comprehensive description of the structural and evolutionary relationships of known structure. Proteins are classified to reflect both structural and evolutionary relatedness. Many levels exist in the hierarchy, but the principal levels are family, superfamily and fold, as described below.

Family (clear evolutionary relationship) : Proteins clustered together into families are clearly evolutionarily related. Generally, this means that pairwise residue identities between the proteins are 30% and greater. However, in some cases similar functions and structures provide definitive evidence of common descent in the absence of high sequence identity; for example, many globins form a family though some members have sequence identities of only 15%.

Superfamily (probable common evolutionary origin : Proteins that have low sequence identities, but whose structural and functional features suggest that a common evolutionary origin is probable, are placed together in superfamilies. For example, actin, the ATPase domain of the heat shock protein, and hexakinase together form a superfamily.

Fold (major structural similarity): Proteins are defined as having a common fold if they have the same major secondary structures in the same arrangement and with the same topological connections. Different proteins with the same fold often have peripheral elements of secondary structure and turn regions that differ in size and conformation. In some cases, these differing peripheral regions may comprise half the structure. Proteins placed together in the same fold category may not have a common evolutionary origin: the structural similarities could arise just from the physics and chemistry of proteins favouring certain packing arrangements and chain topologies.

Method

A general paradigm of protein threading consists of the following four steps:

The construction of a structure template database : Select protein structures from the protein structure databases as structural templates. This generally involves selecting protein structures from databases such as PDB, FSSP, SCOP, or CATH, after removing protein structures with high sequence similarities.

The design of the scoring function : Design a good scoring function to measure the fitness between target sequences and templates based on the knowledge of the known relationships between the structures and the sequences. A good scoring function should contain mutation potential, environment fitness potential, pairwise potential, secondary structure compatibilities, and gap penalties. The quality of the energy function is closely related to the prediction accuracy, especially the alignment accuracy.

Threading alignmen : Align the target sequence with each of the structure templates by optimizing the designed scoring function. This step is one of the major tasks of all threading-based structure prediction programs that take into account the pairwise contact potential; otherwise, a dynamic programming algorithm can fulfil it. This thesis is mainly dedicated to solving the optimal alignment problem derived from a scoring function considering pairwise contacts.

Threading prediction: Select the threading alignment that is statistically most probable as the threading prediction. Then construct a structure model for the target by placing the backbone atoms of the target sequence at their aligned backbone positions of the selected structural template.

Comparison with Homology Modelling

Homology modelling and protein threading are both template-based methods and there is no rigorous boundary between them in terms of prediction techniques. But the protein structures they target are different. Homology modelling is for those targets that have homologous proteins with known structure, while protein threading is for those targets with only fold-level homology found. In other words, homology modelling is for "easier" targets and protein threading is for "harder" targets. Homology modelling treats the template in an alignment as a sequence, and only sequence homology is used for prediction. Protein threading treats the template in an alignment as a structure, and both sequence and structure information extracted from the alignment are used for prediction. When there is no significant homology found, protein threading can make a prediction based on the structure information. That also explains why protein threading may be more effective than homology modelling in many cases.

In practice, when the sequence identity in a sequence sequence alignment is low (i.e. <25%), homology modelling may not produce a significant prediction. In this case, if there is distant homology found for the target, protein threading can generate a good prediction.

More about Threading

Fold recognition methods can be broadly divided into two types: 1, those that derive a 1-D profile for each structure in the fold library and align the target sequence to these profiles; and 2, those that consider the full 3-D structure of the protein template. A simple example of a profile representation would be to take each amino acid in the structure and simply label it according to whether it is buried in the core of the protein or exposed on the surface.

More elaborate profiles might take into account the local secondary structure (e.g. whether the amino acid is part of an alpha helix) or even evolutionary information (how conserved the amino acid is). In the 3-D representation, the structure is modelled as a set of inter-atomic distances, i.e. the distances are calculated between some or all of the atom pairs in the structure. This is a much richer and far more flexible description of the structure, but is much harder to use in calculating an alignment. The profile-based fold recognition approach was first described by Bowie, Luthy and Eisenberg in 1991. The term *threading* was first coined by Jones, Taylor and Thornton in 1992, and originally referred specifically to the use of a full 3-D structure atomic representation of the protein template in fold recognition. Today, the terms threading and fold recognition are frequently (though somewhat incorrectly) used interchangeably.

Fold recognition methods are widely used and effective because it is believed that there are a strictly limited number of different protein folds in nature, mostly as a result of evolution but also due to constraints imposed by the basic physics and chemistry of polypeptide chains. There is, therefore, a good chance (currently 70-80%) that a protein which has a similar fold to the target protein has already been studied by X-ray crystallography or nuclear magnetic resonance (NMR) spectroscopy and can be found in the PDB. Currently there are just over 1100 different protein folds known, but new folds are still being discovered every year due in significant part to the ongoing structural genomics projects. Many different algorithms have been proposed for finding the correct threading of a sequence onto a structure, though many make use of dynamic programming in some form. For full 3-D threading, the problem of identifying the best alignment is very difficult (it is an NP-hard problem for some models of threading). Researchers have made use of many combinatorial optimization methods such as simulated annealing or branch and bound searching to arrive at heuristic solutions.

It is interesting to compare threading methods to methods which attempt to align two protein structures (protein structural alignment), and indeed many of the same algorithms have been applied to both problems.

Protein Threading Software

- HHpred is a popular threading server which runs HHsearch, a widely used software for remote homology detection based on pairwise comparison of hidden Markov models.
- RAPTOR (software) is an integer programming based protein threading software.

Side Chain Geometry Prediction

Accurate packing of the amino acid side chains represents a separate problem. Methods that specifically address the problem of predicting side chain geometry include dead-end elimination and the self-consistent mean field methods. The side chain conformations with low energy are usually determined on the rigid polypeptide backbone and using a set of discrete side chain conformations known as "rotamers" or a "conformational isomerism". The methods attempt to identify the set of rotamers that minimize the model's overall energy.

These methods use rotamer libraries, the collections of rotamers (favourable multi-angle conformations) for each residue type in proteins. Rotamer libraries may contain information about the conformation, its frequency, and the variance about mean dihedral angles, which can be used in sampling. Rotamer libraries are derived from structural bioinformatics or other statistical analysis of side-chain conformations in known experimental structures of proteins, such as by clustering the observed conformations for tetrahedral carbons near the staggered (60°, 180°, -60°) values.

Rotamer libraries can be backbone-independent, secondary-structure-dependent, or backbone-dependent. Backbone-independent rotamer libraries make no reference to backbone conformation, and are calculated from all available side chains of a certain type (for instance, the first example of a rotamer library, done by Ponder and Richards at Yale in 1987). Secondary-structure-dependent libraries present different dihedral angles and/or rotamer frequencies for α-helix, β-sheet, or coil secondary structures.

Backbone-dependent rotamer libraries present conformations and/or frequencies dependent on the local backbone conformation as defined by the backbone dihedral angles ϕ and ψ, regardless of secondary structure. The modern versions of these "libraries" as used in most software are presented as multidimensional distributions of probability or frequency, where the peaks correspond to the dihedral-angle conformations considered as individual rotamers in the lists. Some versions are especially sensitive to the prohibited regions in that conformational space and are used primarily for structure validation, while others emphasize relative frequencies in the favourable regions and are the form used primarily for structure prediction, such as the Dunbrack rotamer "libraries".

The side chain packing methods are most useful for analysing the protein's hydrophobic core, where side chains are more closely packed; they have more difficulty addressing the looser constraints and higher flexibility of surface residues, which often occupy multiple rotamer conformations rather than just one.

Prediction of Structural Classes

Statistical methods have been developed for predicting structural classes of proteins based on their amino acid composition, pseudo amino acid composition and functional domain composition.

Chapter 3

A Theory of Technology Evolution in Animals

Each stage of biological evolution and technological innovation increases the level of biological or technological order and enables entry into the next stage. Order in this case is defined as how well the form (genetic or physical) suits the given function. For example, a modern car is better suited for transporting human occupants than a Model T because the newer vehicle possesses numerous design improvements the older one lacks. An increase in order also generally correlates with an increase in complexity. Getting back to the previous example, a modern car has more moving parts than a Model T, and the newer car's parts are also more precisely and elaborately made. Thus, the modern car is more complex. The overall trend towards increasing complexity can also be observed in biology, where genomes of living species have gradually grown in size along with levels of physical diversity and specialization.

Of course, greater order does not necessarily always entail greater complexity. Sometimes, the simpler solution proves superior both in biology and technology, but the general trends in both have been towards more complexity and more order. The Law of Accelerating Returns states that biological and technological evolution utilize positive feedback, with each improvement building upon the last and enabling the next.

Technological progress in any field consists of a series of "paradigms"—particular methods used to solve certain problems. A given example of a paradigm would be the shrinking of computer transistors to make the computers more powerful. While there are innumerable technological paradigms, all share the same basic life cycles. The advances in the cost-performance of a particular technology, or the Life Cycle of the Paradigm, if graphed, will appear as a sigmoidal S-shape with three distinct phases.

Slow Growth Phase

At this point, the kinks in the technology are still being worked out, and it is still struggling to establish a market base. Growth in price-performance and capabilities is exponential, but still at such an early stage that the growth appears deceptively flat and linear.

Rapid Growth Phase

Begins after the exponential growth passes the "knee of the curve" and explosive growth in the technology's capabilities and user base starts.

Leveling Off Phase

The technology matures as scientists find it increasingly more difficult to make improvements to the same technology in an effort to further address the original need. Growth in usefulness levels off. Once a technology has reached maturity, it is replaced by a newer, totally different technology, meaning a paradigm shift occurs. This occurred during the 1960s when scientists found it impractical to shrink computer vacuum tubes any further and instead switched to transistors, which were newer and allowed the process of miniaturization to continue.

This process of periodic exponential growth parallels biological evolution in two ways. First, biological evolution also occurs in spurts and, second, some biological innovations make organisms exponentially better or speed up evolution from that point onwards in an exponential manner. For example, the advent of DNA allowed life forms to evolve much higher levels of complexity and order.

Society's acceptance of new technologies is speeding up exponentially

- U.S. Phone company revenues and daily number of American phone calls (increasing exponentially)
- Number of U.S. cell phone subscribers (increasing exponentially)

- Time until major new inventions reach mass use by American consumers (decreasing exponentially)

Technologies experiencing exponential changes:

- Increasing exponentially:
    - o Microprocessor clock speeds
    - o Transistors per microprocessor
    - o Processor performance
    - o Dynamic RAM price performance (improving exponentially)
    - o Random Access Memory bits per dollar
    - o Magnetic data storage bits per dollar
    - o Wireless Internet and phone services price performance
    - o Number of Internet hosts
    - o Bytes of Internet traffic
    - o Internet backbone bandwidth (increasing in a very terraced, quasi-exponential manner)
    - o Number of scientific citations for nanotechnology research
    - o Number of U.S. nanotech patents
- Decreasing exponentially:
    - o Average Transistor price
    - o Transistor Manufacturing costs
    - o Dynamic RAM size (smallest feature sizes decreasing exponentially)
    - o Microprocessor costs
    - o DNA sequencing costs per base pair
    - o Mechanical device sizes

While Gordon Moore first observed in 1965 that the transistor densities of integrated circuits were doubling every two years, an extended analysis shows that computers have been experiencing exponential improvements to their cost-performance (maximum number of calculations per second per $1,000) since at least 1900, when the very first electromechanical computers were invented.

This trend in increasing performance has held steady across five computer paradigm shifts (electromechanical, relay-based, vacuum tube, transistor, and integrated circuit) and is now encapsulated by Moore's Law. While integrated circuits will—like all paradigms—ultimately reach the limits of their possible capabilities, the exponentially growing performance trend will likely continue via a paradigm shift to a newer technology like memristors or three-dimensional molecular computing. Extrapolating this rate of improvement, supercomputers should be capable of the same number of calculations per second as a human brain by 2010, and personal computers should be at this level around the year 2020. The amount of improvement in nanoscience at the same rate also see 2020 as a nexus point.

Predictions

2010

- Supercomputers will have the same raw power as human brains (although not yet the equivalently flexible software).
- Computers will disappear as distinct physical objects, being embedded in clothing and in everyday objects.
- Displays built into our eyeglasses will provide full-immersion audio-visual virtual reality.

2010s

- More and more computer devices will be used as miniature web servers, and more will have their resources pooled for computation.
- High-quality broadband Internet access will become available almost everywhere.
- Eyeglasses that beam images onto the users' retinas to produce virtual reality will be developed. They will also come with speakers or headphone attachments that will complete the experience with sounds. These eyeglasses will become a new medium for advertising as advertising will be wirelessly transmitted to them as one walks by various business establishments.
- The VR glasses will also have built-in computers featuring "virtual assistant" programs that can help the user with various daily tasks.

- Virtual assistants would be capable of multiple functions. One useful function would be real-time language translation in which words spoken in a foreign language would be translated into text that would appear as subtitles to a user wearing the glasses.
- Cell phones will be built into clothing and will be able to project sounds directly into the ears of their users.
- Advertisements will utilize a new technology whereby two ultrasonic beams can be targeted to intersect at a specific point, delivering a localized sound message that only a single person can hear. This was fictionalized in the films Minority Report and Back to the Future Part II. See Sound from ultrasound.

2014

- Automatic house cleaning robots will have become common.

2018

- 10 Terabits (10 bits) of computer memory—roughly the equivalent of the memory (RAM) space in a single human brain—will cost $1000.

2020

- Personal computers will have the same processing power as human brains.

2020s

- Computers less than 100 nm in size will be possible.
- As one of their first practical applications, nanomachines will be used for medical purposes.
- Highly advanced medical nanobots will perform detailed brainscans on live patients.
- Accurate computer simulations of the entire human brain will exist due to these hyperaccurate brainscans, and the workings of the brain will be understood.
- Nanobots capable of entering the bloodstream to "feed" cells and extract waste will exist (though not necessarily be in wide use) by the end of this decade. They will make the normal mode of human food consumption

obsolete. Thus, humans who have injected these nanobots into their bloodstream will evolve from having a normal human metabolism and become humanoid cyborgs. Eventually, according to Kurzweil, a large percentage of humans will evolve by this process into cyborgs.

- By the late 2020s, nanotech-based manufacturing will be in widespread use, radically altering the economy as all sorts of products can suddenly be produced for a fraction of their traditional-manufacture costs. The true cost of any product is now the amount of time it takes to download the design schematics.
- Also by the later part of this decade, virtual reality will be so high-quality that it will be indistinguishable from reality.
- The threat posed by genetically engineered pathogens permanently dissipates by the end of this decade as medical nanobots—far more durable, intelligent and capable than any microorganism—become sufficiently advanced.
- A computer will pass the Turing test by the last year of the decade (2029), meaning that it is a Strong AI and can think like a human (though the first A.I. is likely to be the equivalent of a kindergartner). This first A.I. is built around a computer simulation of a human brain, which was made possible by previous, nanotech-guided brainscanning.

2025

- The most likely year for the debut of advanced nanotechnology.
- Some military UAVs and land vehicles will be 100% computer-controlled.

2030s

- Mind uploading becomes possible.
- Nanomachines could be directly inserted into the brain and could interact with brain cells to totally control incoming and outgoing signals. As a result, truly full-immersion virtual reality could be generated without

the need for any external equipment. Afferent nerve pathways could be blocked, totally canceling out the real world and leaving the user with only the desired virtual experience.

- Brain nanobots could also elicit emotional responses from users.
- Using brain nanobots, recorded or real-time brain transmissions
  of a person's daily life known as "experience beamers" will be available for other people to remotely experience. This is very similar to how the characters in Being John Malkovich were able to enter the mind of Malkovich and see the world through his eyes.
- Recreational uses aside, nanomachines in people's brains will allow them to greatly expand their cognitive, memory and sensory capabilities, to directly interface with computers, and to telepathically communicate with other, similarly augmented humans via wireless networks.
- The economy transits in GDP percentage to more meta services such as reality fabrication, mind enhancement, mental software.
- The same nanotechnology should also allow people to alter the neural connections within their brains, changing the underlying basis for the person's intelligence, memories and personality.

2040s

- Human body 3.0 (as Kurzweil calls it) comes into existence. It lacks a fixed, corporeal form and can alter its shape and external appearance at will via foglet-like nanotechnology. Organs are also replaced by superior cybernetic implants.
- There will be social splitting into different levels of use of reality augmentation, from those who want to live in a life of imagined harems, or those who dedicate their thoughts to philosophical extension. Human society will drift apart in its focus, but with ever increasing capabilities to make imagined things occur.

- People spend most of their time in full-immersion virtual reality (Kurzweil has cited The Matrix as a good example of what the advanced virtual worlds will be like, without the dystopian twist).
- Foglets are in use.

2045: The Singularity

- $1000 buys a computer a billion times more powerful than the human brain. This means that average and even low-end computers are hugely smarter than even highly intelligent, unenhanced humans.
- The Singularity occurs as artificial intelligences surpass human beings as the smartest and most capable life forms on the Earth. Technological development is taken over by the machines, who can think, act and communicate so quickly that normal humans cannot even comprehend what is going on; thus the machines, acting in concert with those humans who have evolved into postbiological cyborgs, achieve effective world domination. The machines enter into a "runaway reaction" of self-improvement cycles, with each new generation of A.I.s appearing faster and faster. From this point onwards, technological advancement is explosive, under the control of the machines, and thus cannot be accurately predicted.
- The Singularity is an extremely disruptive, world-altering event that forever changes the course of human history. The extermination of humanity by violent machines is unlikely (though not impossible) because sharp distinctions between man and machine will no longer exist thanks to the existence of cybernetically enhanced humans and uploaded humans.

Post-2045: "Waking up" the Universe

- The physical bottom limit to how small computer transistors can be shrunk is reached. From this moment onwards, computers can only be made more powerful if they are made larger in size.
- Because of this, A.I.s convert more and more of the Earth's matter into engineered, computational substrate

capable of supporting more A.I.s. until the whole Earth is one, gigantic computer (but some areas will remain set aside as nature preserves).

- At this point, the only possible way to increase the intelligence of the machines any further is to begin converting all of the matter and energy in the universe into similar massive computers. A.I.s radiate out into space in all directions from the Earth, breaking down whole planets, stars, moons and meteoroids and reassembling them into giant computers. This, in effect, "wakes up" the universe as all the inanimate "dumb" matter (rocks, dust, gases, etc.) is converted into structured matter capable of supporting intelligence, or a form of synthetic life.
- Kurzweil predicts that machines will have the ability to make planet-sized computers by 2099, which underscores how enormously technology will advance after the Singularity.
- The process of "waking up" the universe will be complete as early as 2199.
- With the entire universe made into a giant, highly efficient supercomputer, A.I./human hybrids (so integrated that, in truth it is a new category of "life") would have both supreme intelligence and physical control over the universe. Kurzweil suggests that this will open up all sorts of new possibilities, including manipulation of the physical constants, inter-dimensional travel, and controlling the fate of the universe.

Some indeterminate point within a few decades from now

- Space technology becomes advanced enough to provide the Earth permanent protection from the threat of asteroid impacts.
- The antitechnology "Luddite" movement will grow increasingly vocal and possibly resort to violence, possibly a new World War, as these people become enraged over the emergence of new technologies that threaten traditional attitudes regarding the nature of human life (radical life extension, genetic engineering,

cybernetics) and the supremacy of mankind (artificial intelligence). Though the Luddites might, at best, succeed in delaying the Singularity, the march of technology is irresistible and they will inevitably fail in keeping the world frozen at a fixed level of development. However, some nature preserves may be set aside for them to live in.

- The emergence of distributed energy grids and full-immersion virtual reality will, when combined with high bandwidth Internet, enable the ultimate in telecommuting. This, in turn, will make cities obsolete since workers will no longer need to be located near their workplaces. The decentralization of the population will make societies less vulnerable to terrorist and military attacks.

Film Adaptations

In 2006, Barry Ptolemy and his production company Ptolemaic Productions licensed the rights to The Singularity Is Near from Kurzweil. Inspired by the book, Ptolemy would direct and produce the film Transcendent Man, which would go on to critical and commercial success in 2011, bringing more attention to the book. Kurzweil considers Doug Wolen's coming documentary "The Singularity" to be closer to form to Kurzweil's own views

Translations

- Chinese translation
- Italian translation
- Korean translation
- DNA microarray

A DNA microarray (also commonly known as gene chip, DNA chip, or biochip) is a collection of microscopic DNA spots attached to a solid surface. Scientists use DNA microarrays to measure the expression levels of large numbers of genes simultaneously or to genotype multiple regions of a genome.

Each DNA spot contains picomoles (10 moles) of a specific DNA sequence, known as probes (or reporters). These can be a short section of a gene or other DNA element that are used to hybridize a cDNA or cRNA sample (called target) under high-stringency conditions. Probe-target hybridization is usually detected and quantified by detection of fluorophore-, silver-, or chemiluminescence-labeled targets to determine relative abundance of nucleic acid sequences in the target.

The Basic Microarray

Since an array can contain tens of thousands of probes, a microarray experiment can accomplish many genetic tests in parallel. Therefore arrays have dramatically accelerated many types of investigation. In standard microarrays, the probes are synthesized and then attached via surface engineering to a solid surface by a covalent bond to a chemical matrix (via epoxy-silane, amino-silane, lysine, polyacrylamide or others). The solid surface can be glass or a silicon chip, in which case they are colloquially known as an Affy chip when an Affymetrix chip is used. Other microarray platforms, such as Illumina, use microscopic beads, instead of the large solid support. Alternatively, microarrays can be constructed by the direct synthesis of oligonucleotide probes on solid surfaces. DNA arrays are different from other types of microarray only in that they either measure DNA or use DNA as part of its detection system.

DNA microarrays can be used to measure changes in expression levels, to detect single nucleotide polymorphisms (SNPs), or to genotype or resequence mutant genomes. Microarrays also differ in fabrication, workings, accuracy, efficiency, and cost. Additional factors for microarray experiments are the experimental design and the methods of analyzing the data.

History

Microarray technology evolved from Southern blotting, where fragmented DNA is attached to a substrate and then probed with a known gene or fragment. The first reported use of this approach was the analysis of 378 arrayed lysed bacterial colonies each harboring a different sequence which were assayed in multiple replicas for expression of the genes in multiple normal and tumor tissue. This was expanded to analysis of more than 4000 human sequences with computer driven scanning and image processing for quantitative analysis of the sequences in human colonic tumors and normal tissue and then to comparison of colonic tissues at different genetic risk. The use of a collection of distinct DNAs in arrays for expression profiling was also described in 1987, and the arrayed DNAs were used to identify genes whose expression is modulated by interferon. These early gene arrays were made by spotting cDNAs onto filter paper with a pin-spotting device. The use of miniaturized microarrays for gene expression profiling was first reported in 1995, and a complete eukaryotic genome (Saccharomyces cerevisiae) on a microarray was published in 1997.

Principle

Hybridization of the Target to the Probe

The core principle behind microarrays is hybridization between two DNA strands, the property of complementary nucleic acid sequences to specifically pair with each other by forming hydrogen bonds between complementary nucleotide base pairs. A high number of complementary base pairs in a nucleotide sequence means tighter non-covalent bonding between the two strands. After washing off of non-specific bonding sequences, only strongly paired strands will remain hybridized. So fluorescently labeled target sequences that bind to a probe sequence generate a signal that depends on the strength of the hybridization determined by the number of paired bases, the hybridization conditions (such as temperature), and washing after hybridization. Total strength of the signal, from a spot (feature), depends upon the amount of target sample binding to the probes present on that spot. Microarrays use relative quantization in which the intensity of a feature is compared to the intensity of the same feature under a different condition, and the identity of the feature is known by its position.

Uses and Types

Many types of array exist and the broadest distinction is whether they are spatially arranged on a surface or on coded beads:

- The traditional solid-phase array is a collection of orderly microscopic "spots", called features, each with a specific probe attached to a solid surface, such as glass, plastic or silicon biochip (commonly known as a genome chip,

DNA chip or gene array). Thousands of them can be placed in known locations on a single DNA microarray.

- The alternative bead array is a collection of microscopic polystyrene beads, each with a specific probe and a ratio of two or more dyes, which do not interfere with the fluorescent dyes used on the target sequence.

DNA microarrays can be used to detect DNA (as in comparative genomic hybridization), or detect RNA (most commonly as cDNA after reverse transcription) that may or may not be translated into proteins. The process of measuring gene expression via cDNA is called expression analysis or expression profiling.

Fabrication

Microarrays can be manufactured in different ways, depending on the number of probes under examination, costs, customization requirements, and the type of scientific question being asked. Arrays may have as few as 10 probes or up to 2.1 million micrometre-scale probes from commercial vendors.

Spotted vs. in Situ Synthesised Arrays

Microarrays can be fabricated using a variety of technologies, including printing with fine-pointed pins onto glass slides, photolithography using pre-made masks, photolithography using dynamic micromirror devices, ink-jet printing, or electrochemistry on microelectrode arrays. In spotted microarrays, the probes are oligonucleotides, cDNA or small fragments of PCR products that correspond to mRNAs. The probes are synthesized prior to deposition on the array surface and are then “spotted” onto glass. A common approach utilizes an array of fine pins or needles controlled by a robotic arm that is dipped into wells containing DNA probes and then depositing each probe at designated locations on the array surface.

The resulting “grid” of probes represents the nucleic acid profiles of the prepared probes and is ready to receive complementary cDNA or cRNA “targets” derived from experimental or clinical samples. This technique is used by research scientists around the world to produce “in-house” printed microarrays from their own labs. These arrays may be easily customized for each experiment, because researchers can choose the probes and printing locations on the arrays, synthesize the probes in their own lab (or collaborating facility), and spot the arrays.

They can then generate their own labeled samples for hybridization, hybridize the samples to the array, and finally scan the arrays with their own equipment. This provides a relatively low-cost microarray that may be customized for each study, and avoids the costs of purchasing often more expensive commercial arrays that may represent vast numbers of genes that are not of interest to the investigator. Publications exist which indicate in-house spotted microarrays may not provide the same level of sensitivity compared to commercial oligonucleotide arrays, possibly owing to the small batch sizes and reduced printing efficiencies when compared to industrial manufactures of oligo arrays.

In oligonucleotide microarrays, the probes are short sequences designed to match parts of the sequence of known or predicted open reading frames. Although oligonucleotide probes are often used in "spotted" microarrays, the term "oligonucleotide array" most often refers to a specific technique of manufacturing. Oligonucleotide arrays are produced by printing short oligonucleotide sequences designed to represent a single gene or family of gene splice-variants by synthesizing this sequence directly onto the array surface instead of depositing intact sequences. Sequences may be longer (60-mer probes such as the Agilent design) or shorter (25-mer probes produced by Affymetrix) depending on the desired purpose; longer probes are more specific to individual target genes, shorter probes may be spotted in higher density across the array and are cheaper to manufacture. One technique used to produce oligonucleotide arrays include photolithographic synthesis (Affymetrix) on a silica substrate where light and light-sensitive masking agents are used to "build" a sequence one nucleotide at a time across the entire array.

Each applicable probe is selectively "unmasked" prior to bathing the array in a solution of a single nucleotide, then a masking reaction takes place and the next set of probes are unmasked in preparation for a different nucleotide exposure. After many repetitions, the sequences of every probe become fully constructed. More recently, Maskless Array Synthesis from NimbleGen Systems has combined flexibility with large numbers of probes.

Two-channel vs. One-channel Detection

Two-color microarrays or two-channel microarrays are typically hybridized with cDNA prepared from two samples to be compared (e.g. diseased tissue versus healthy tissue) and that are labeled with two different fluorophores. Fluorescent dyes commonly used for cDNA labeling include Cy3, which has a fluorescence emission wavelength of 570 nm (corresponding to the green part of the light spectrum), and Cy5 with a fluorescence emission wavelength of 670 nm (corresponding to the red part of the light spectrum). The two Cy-labeled cDNA samples are mixed and hybridized to a single microarray that is then scanned in a microarray scanner to visualize fluorescence of the two fluorophores after excitation with a laser beam of a defined wavelength. Relative intensities of each fluorophore may then be used in ratio-based analysis to identify up-regulated and down-regulated genes.

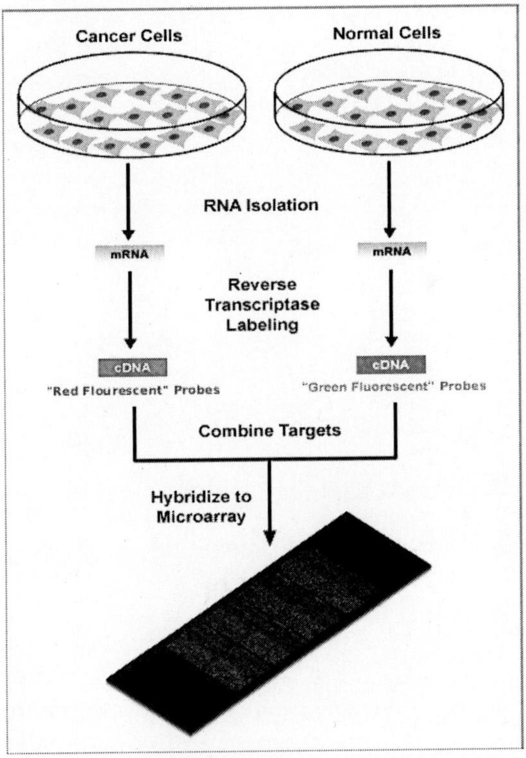

***Figure:** Diagram of typical dual-colour microarray experiment.*

Oligonucleotide microarrays often carry control probes designed to hybridize with RNA spike-ins. The degree of hybridization between the spike-ins and the control probes is used to normalize the hybridization measurements for the target probes. Although absolute levels of gene expression may be determined in the two-color array in rare instances, the relative differences in expression among different spots within a sample and between samples is the preferred method of data analysis for the two-color system. Examples of providers for such microarrays includes Agilent with their Dual-Mode platform, Eppendorf with their DualChip platform for colorimetric Silverquant labeling, and TeleChem International with Arrayit.

In single-channel microarrays or one-color microarrays, the arrays provide intensity data for each probe or probe set indicating a relative level of hybridization with the labeled target. However, they do not truly indicate abundance levels of a gene but rather relative abundance when compared to other samples or conditions when processed in the same experiment. Each RNA molecule encounters protocol and batch-specific bias during amplification, labeling, and hybridization phases of the experiment making comparisons between genes for the same microarray uninformative. The comparison of two conditions for the same gene requires two separate single-dye hybridizations.

Several popular single-channel systems are the Affymetrix "Gene Chip", Illumina "Bead Chip", Agilent single-channel arrays, the Applied Microarrays "CodeLink" arrays, and the Eppendorf "DualChip & Silverquant". One strength of the single-dye system lies in the fact that an aberrant sample cannot affect the raw data derived from other samples, because each array chip is exposed to only one sample (as opposed to a two-color system in which a single low-quality sample may drastically impinge on overall data precision even if the other sample was of high quality). Another benefit is that data are more easily compared to arrays from different experiments so long as batch effects have been accounted for. A drawback to the one-color system is that, when compared to the two-color system, twice as many microarrays are needed to compare samples within an experiment.

Microarrays and Bioinformatics

Gene expression values from microarray experiments can be represented as heat maps to visualize the result of data analysis. The advent of inexpensive microarray experiments created several specific bioinformatics challenges:

- the multiple levels of replication in experimental design (Experimental design)
- the number of platforms and independent groups and data format (Standardization)
- the treatment of the data (Statistical analysis)
- accuracy and precision (Relation between probe and gene)
- the sheer volume of data and the ability to share it (Data warehousing)

Experimental Design

Due to the biological complexity of gene expression, the considerations of experimental design that are discussed in the expression profiling article are of critical importance if statistically and biologically valid conclusions are to be drawn from the data. There are three main elements to consider when designing a microarray experiment. First, replication of the biological samples is essential for drawing conclusions from the experiment. Second, technical replicates (two RNA samples obtained from each experimental unit) help to ensure precision and allow for testing differences within treatment groups. The biological replicates include independent RNA extractions and technical replicates may be two aliquots of the same extraction.

Third, spots of each cDNA clone or oligonucleotide are present as replicates (at least duplicates) on the microarray slide, to provide a measure of technical precision in each hybridization. It is critical that information about the sample preparation and handling is discussed, in order to help identify the independent units in the experiment and to avoid inflated estimates of statistical significance.

Standardization

Microarray data is difficult to exchange due to the lack of standardization in platform fabrication, assay protocols, and analysis methods. This presents an interoperability problem in bioinformatics. Various grass-roots open-source projects are trying to ease the exchange and analysis of data produced with non-proprietary chips:

- For example, the "Minimum Information About a Microarray Experiment" (MIAME) checklist helps define the level of detail that should exist and is being adopted by many journals as a requirement for the submission of papers incorporating microarray results. But MIAME does not describe the format for the information, so while many formats can support the MIAME requirements, as of 2007 no format permits verification of complete semantic compliance.
- The "MicroArray Quality Control (MAQC) Project" is being conducted by the US Food and Drug Administration (FDA) to develop standards and quality control metrics which will eventually allow the use of MicroArray data in drug discovery, clinical practice and regulatory decision-making.
- The MGED Society has developed standards for the representation of gene expression experiment results and relevant annotations.

Statistical Analysis

Microarray data sets are commonly very large, and analytical precision is influenced by a number of variables. Statistical challenges include taking into account effects of background noise and appropriate normalization of the data. Normalization methods may be suited to specific platforms and, in the case of commercial platforms, the analysis may be proprietary. Algorithms that affect statistical analysis include:

- Image analysis: gridding, spot recognition of the scanned image (segmentation algorithm), removal or marking of poor-quality and low-intensity features (called flagging).
- Data processing: background subtraction (based on global or local background), determination of spot intensities and intensity ratios, visualisation of data, and log-transformation of ratios, global or local normalization of intensity ratios, and segmentation into different copy number regions using step detection algorithms.

- Identification of statistically significant changes: t-test, ANOVA, Bayesian method Mann–Whitney test methods tailored to microarray data sets, which take into account multiple comparisons or cluster analysis. These methods assess statistical power based on the variation present in the data and the number of experimental replicates, and can help minimize Type I and type II errors in the analyses.
- Network-based methods: Statistical methods that take the underlying structure of gene networks into account, representing either associative or causative interactions or dependencies among gene products.

Microarray data may require further processing aimed at reducing the dimensionality of the data to aid comprehension and more focused analysis. Other methods permit analysis of data consisting of a low number of biological or technical replicates; for example, the Local Pooled Error (LPE) test pools standard deviations of genes with similar expression levels in an effort to compensate for insufficient replication.

Relation Between Probe and Gene

The relation between a probe and the mRNA that it is expected to detect is not trivial. Some mRNAs may cross-hybridize probes in the array that are supposed to detect another mRNA. In addition, mRNAs may experience amplification bias that is sequence or molecule-specific. Thirdly, probes that are designed to detect the mRNA of a particular gene may be relying on genomic EST information that is incorrectly associated with that gene.

Data Warehousing

Microarray data was found to be more useful when compared to other similar datasets. The sheer volume of data, specialized formats (such as MIAME), and curation efforts associated with the datasets require specialized databases to store the data.

DNA Sequencing

DNA sequencing includes several methods and technologies that are used for determining the order of the nucleotide bases—adenine, guanine, cytosine, and thymine—in a molecule of DNA. Knowledge of DNA sequences has become indispensable for basic biological research, other research branches utilizing DNA sequencing, and in numerous applied fields such as diagnostic, biotechnology, forensic biology and biological systematics. The advent of DNA sequencing has significantly accelerated biological research and discovery. The rapid speed of sequencing attained with modern DNA sequencing technology has been instrumental in the sequencing of the human genome, in the Human Genome Project. Related projects, often by scientific collaboration across continents, have generated the complete DNA sequences of many animal, plant, and microbial genomes.

The first DNA sequences were obtained in the early 1970s by academic researchers using laborious methods based on two-dimensional chromatography. Following the development of dye-based sequencing methods with automated analysis, DNA sequencing has become easier and orders of magnitude faster.

History

RNA sequencing was one of the earliest forms of nucleotide sequencing. The major landmark of RNA sequencing is the sequence of the first complete gene and the complete genome of Bacteriophage MS2, identified and published by Walter Fiers and his coworkers at the University of Ghent (Ghent, Belgium), between 1972 and 1976. Prior to the development of rapid DNA sequencing methods in the early 1970s by Frederick Sanger at the University of Cambridge, in England and Walter Gilbert and Allan Maxam at Harvard, a number of laborious methods were used. For instance, in 1973, Gilbert and Maxam reported the sequence of 24 basepairs using a method known as wandering-spot analysis. The chain-termination method developed by Sanger and coworkers in 1977 soon became the method of choice, owing to its relative ease and reliability. It involves separating DNA bases from different DNA fragments.

Maxam–Gilbert Sequencing

In 1976–1977, Allan Maxam and Walter Gilbert developed a DNA sequencing method based on chemical modification of DNA and subsequent cleavage at specific bases. Although Maxam and Gilbert published their chemical sequencing method two years after the ground-breaking paper of Sanger and Coulson on plus-minus sequencing, Maxam–Gilbert sequencing rapidly became more popular, since purified DNA could be used directly, while the initial Sanger method required that each read start be cloned for production of single-stranded DNA.

However, with the improvement of the chain-termination method, Maxam-Gilbert sequencing has fallen out of favour due to its technical complexity prohibiting its use in standard molecular biology kits, extensive use of hazardous chemicals, and difficulties with scale-up. The method requires radioactive labeling at one 5' end of the DNA (typically by a kinase reaction using gamma- P ATP) and purification of the DNA fragment to be sequenced. Chemical treatment generates breaks at a small proportion of one or two of the four nucleotide bases in each of four reactions (G, A+G, C, C+T). For example, the purines (A+G) are depurinated using formic acid, the guanines (and to some extent the adenines) are methylated by dimethyl sulfate, and the pyrimidines (C+T) are methylated using hydrazine.

The addition of salt (sodium chloride) to the hydrazine reaction inhibits the methylation of thymine for the C-only reaction. The modified DNAs are then cleaved by hot piperidine at the position of the modified base. The concentration of the modifying chemicals is controlled to introduce on average one modification per DNA molecule. Thus a series of labeled fragments is generated, from the radiolabeled end to the first "cut" site in each molecule. The fragments in the four reactions are electrophoresed side by side in denaturing acrylamide gels for size separation. To visualize the fragments, the gel is exposed to X-ray film for autoradiography, yielding a series of dark bands each corresponding to a radiolabeled DNA fragment, from which the sequence may be inferred. Also sometimes known as "chemical sequencing", this method led to the Methylation Interference Assay used to map DNA-binding sites for DNA-binding proteins.

Chain-termination Methods

Because the chain-terminator method (or Sanger method after its developer Frederick Sanger) is more efficient and uses fewer toxic chemicals and lower amounts of radioactivity than the method of Maxam and Gilbert, it rapidly became the method of choice. The key principle of the Sanger method was the use of dideoxynucleotide triphosphates (ddNTPs) as DNA chain terminators.

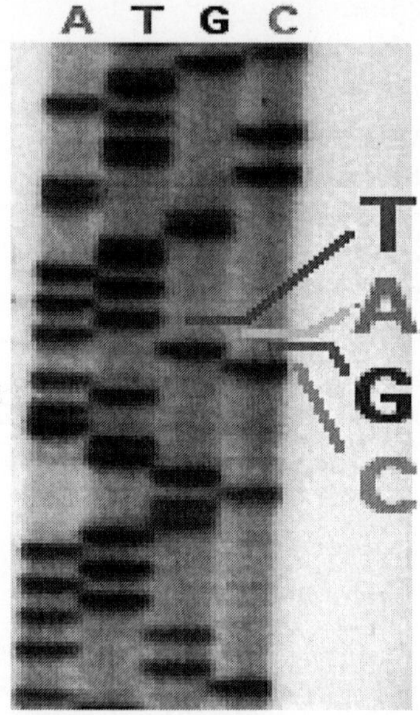

Figure: *Part of a radioactively labelled sequencing gel*

The classical chain-termination method requires a single-stranded DNA template, a DNA primer, a DNA polymerase, normal deoxynucleotidetriphosphates (dNTPs), and modified nucleotides (dideoxyNTPs) that terminate DNA strand elongation. These ddNTPs will also be radioactively or fluorescently labelled for detection in automated sequencing machines.

The DNA sample is divided into four separate sequencing reactions, containing all four of the standard deoxynucleotides (dATP, dGTP, dCTP and dTTP) and the DNA polymerase. To each reaction is added only one of the four dideoxynucleotides (ddATP, ddGTP, ddCTP, or ddTTP) which are the chain-terminating nucleotides, lacking a 3'-OH group required for the formation of a phosphodiester bond between two nucleotides, thus terminating DNA strand extension and resulting in DNA fragments of varying length. The newly synthesized and labelled DNA fragments are heat denatured, and separated by size (with a resolution of just one nucleotide) by gel electrophoresis on a denaturing polyacrylamide-urea gel with each of the four reactions run in one of four individual lanes (lanes A, T, G, C); the DNA bands are then visualized by autoradiography or UV light, and the DNA sequence can be directly read off the X-ray film or gel image. In the image on the right, X-ray film was exposed to the gel, and the dark bands correspond to DNA fragments of different lengths. A dark band in a lane indicates a DNA fragment that is the result of chain termination after incorporation of a dideoxynucleotide (ddATP, ddGTP, ddCTP, or ddTTP). The relative positions of the different bands among the four lanes are then used to read (from bottom to top) the DNA sequence. DNA fragments are labelled with a radioactive or fluorescent tag on the primer (1), in the new DNA strand with a labeled dNTP, or with a labeled ddNTP. (click to expand)

Technical variations of chain-termination sequencing include tagging with nucleotides containing radioactive phosphorus for radiolabelling, or using a primer labeled at the 5' end with a fluorescent dye. Dye-primer sequencing facilitates reading in an optical system for faster and more economical analysis and automation. The later development by Leroy Hood and coworkers of fluorescently labeled ddNTPs and primers set the stage for automated, high-throughput DNA sequencing. Sequence ladder by radioactive sequencing compared to fluorescent peaks (click to expand)

Chain-termination methods have greatly simplified DNA sequencing. For example, chain-termination-based kits are commercially available that contain the reagents needed for sequencing, pre-aliquoted and ready to use. Limitations include non-specific binding of the primer to the DNA, affecting accurate read-out of the DNA sequence, and DNA secondary structures affecting the fidelity of the sequence.

Dye-terminator Sequencing

Dye-terminator sequencing utilizes labelling of the chain terminator ddNTPs, which permits sequencing in a single reaction, rather than four reactions as in the labelled-primer method. In dye-terminator sequencing, each of the four dideoxynucleotide chain terminators is labelled with fluorescent dyes, each of which emit light at different wavelengths.

Owing to its greater expediency and speed, dye-terminator sequencing is now the mainstay in automated sequencing. Its limitations include dye effects due to differences in the incorporation of the dye-labelled chain terminators into the

DNA fragment, resulting in unequal peak heights and shapes in the electronic DNA sequence trace chromatogram after capillary electrophoresis.

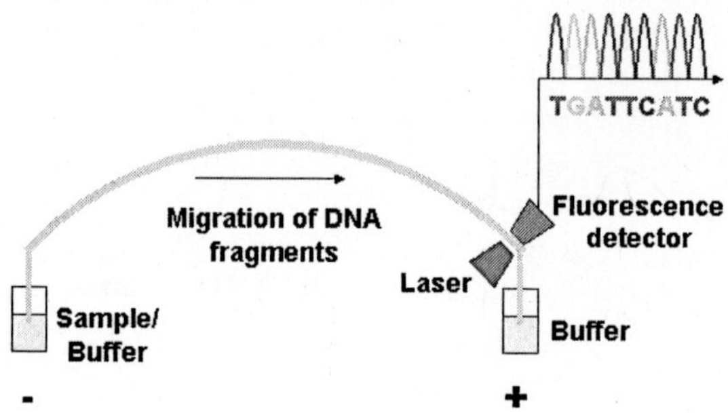

Figure: *Capillary electrophoresis (click to expand)*

This problem has been addressed with the use of modified DNA polymerase enzyme systems and dyes that minimize incorporation variability, as well as methods for eliminating "dye blobs". The dye-terminator sequencing method, along with automated high-throughput DNA sequence analyzers, is now being used for the vast majority of sequencing projects.

Challenges

Common challenges of DNA sequencing include poor quality in the first 15–40 bases of the sequence and deteriorating quality of sequencing traces after 700–900 bases. Base calling software typically gives an estimate of quality to aid in quality trimming. In cases where DNA fragments are cloned before sequencing, the resulting sequence may contain parts of the cloning vector. In contrast, PCR-based cloning and emerging sequencing technologies based on pyrosequencing often avoid using cloning vectors. Recently, one-step Sanger sequencing (combined amplification and sequencing) methods such as Ampliseq and SeqSharp have been developed that allow rapid sequencing of target genes without cloning or prior amplification.

Current methods can directly sequence only relatively short (300–1000 nucleotides long) DNA fragments in a single reaction. The main obstacle to sequencing DNA fragments above this size limit is insufficient power of separation for resolving large DNA fragments that differ in length by only one nucleotide. In all cases the use of a primer with a free 3' end is essential. Automated DNA-sequencing instruments (DNA sequencers) can sequence up to 384 DNA samples in a single batch (run) in up to 24 runs a day. DNA sequencers carry out capillary electrophoresis for size separation, detection and recording of dye fluorescence, and data output as fluorescent peak trace chromatograms. Sequencing reactions by thermocycling, cleanup and re-suspension in a buffer solution before loading onto the sequencer are performed separately. A number of commercial and non-commercial software packages can trim low-quality DNA traces automatically. These programs score the quality of each peak and remove low-quality base peaks (generally located at the ends of the sequence). The accuracy of such algorithms is below visual examination by a human operator, but sufficient for automated processing of large sequence data sets.

Amplification and clonal selection

Large-scale sequencing often aims at sequencing very long DNA pieces, such as whole chromosomes, although large-scale sequencing can also be used to generate very large numbers of short sequences, such as found in phage display. For longer targets, such as chromosomes, common approaches consist of cutting (with restriction enzymes) or shearing (with mechanical forces) large DNA fragments into shorter DNA fragments. The fragmented DNA is cloned into a DNA vector, and amplified in Escherichia coli. Short DNA fragments purified from individual bacterial colonies are individually sequenced and assembled electronically into one long, contiguous sequence.

This method does not require any pre-existing information about the sequence of the DNA and is referred to as de novo sequencing. Gaps in the assembled sequence may be filled by primer walking. The different strategies have different tradeoffs in speed and accuracy; shotgun methods are often used for sequencing large genomes, but its assembly is complex and difficult, particularly with sequence repeats often causing gaps in genome assembly.

Most sequencing approaches use an in vitro cloning step to amplify individual DNA molecules, because their molecular detection methods are not sensitive enough for single molecule sequencing. Emulsion PCR isolates individual DNA molecules along with primer-coated beads in aqueous droplets within an oil phase. Polymerase chain reaction (PCR) then coats each bead with clonal copies of the DNA molecule followed by immobilization for later sequencing. Emulsion PCR is used in the methods by Marguilis et al. (commercialized by 454 Life Sciences), Shendure and Porreca et al. (also known as "Polony sequencing") and SOLiD sequencing, (developed by Agencourt, now Applied Biosystems).

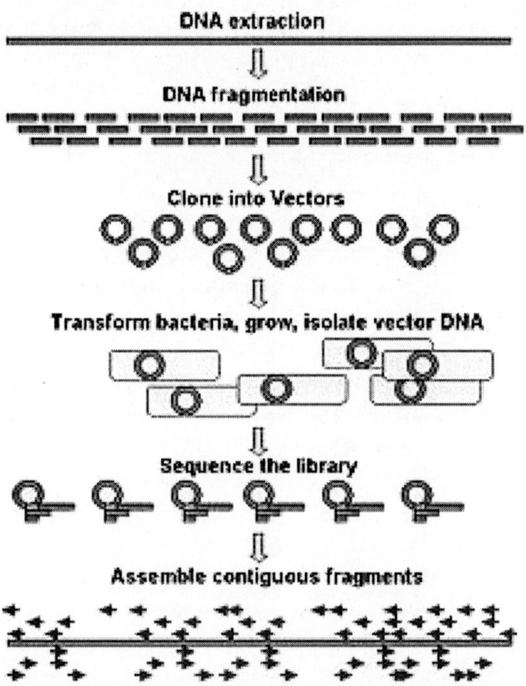

Figure: *Genomic DNA is fragmented into random pieces and cloned as a bacterial library. DNA from individual bacterial clones is sequenced and the sequence is assembled by using overlapping DNA regions.*

Another method for in vitro clonal amplification is bridge PCR, where fragments are amplified upon primers attached to a solid surface, used in the Illumina Genome Analyzer. Single-molecule methods, such as that developed by Stephen Quake's laboratory (later commercialized by Helicos) is an exception: it uses bright fluorophores and laser excitation to detect base addition events from individual DNA molecules fixed to a surface, eliminating the need for molecular amplification.

High-throughput Sequencing

The high demand for low-cost sequencing has driven the development of high-throughput sequencing technologies that parallelize the sequencing process, producing thousands or millions of sequences at once. High-throughput sequencing technologies are intended to lower the cost of DNA sequencing beyond what is possible with standard dye-terminator methods. Massively parallel signature sequencing Massive parallel signature sequencing (MPSS) is a sequenced based approach that can be used to identify and quantify mRNA transcripts present in a sample similar to serial analysis of gene expression (SAGE) but the biochemical manipulation and sequencing approach differ substantially. MPSS allows mRNA transcripts to be identified through the generation of a 17-20 bp (base pair) signature sequence adjacent to the 3'-end of the 3'- most site of the designated restriction enzyme (commonly Sau3A or DpnII). Each signature sequence is cloned onto one of a million microbeads. The technique ensures that only one type of DNA sequence is on a microbead.

So if there are 50 copies of a specific transcript in the biological sample, these transcripts will be captured onto 50 different microbeads, each bead holding roughly 100 000 amplified copies of the specific signature. The microbeads are then arrayed in a flow cell for sequencing and quantification. The sequence signatures are deciphered by the parallel identification of four bases by hybridization to fluorescently labeled encoders. Each of the encoders has a unique label which is detected after hybridization by taking an image of the microbead array. The next step is to cleave and remove that set of four bases and reveal the next four bases for a new round of hybridization to encoders and image acquisition.

The raw output is a list of 17-20 bp signature sequences, that can be annotated to the human genome for gene identification. The longer tag sequence confers a higher specificity than the classical SAGE tag of 9-10 bp. The level of unique gene expression is represented by the count of transcripts present per million molecules, similar to SAGE output. A significant advantage is the larger library size compared with SAGE. An MPSS library typically holds 1 million signature tags, which is roughly 20 times the size of a SAGE library. Some of the disadvantages related to SAGE apply to MPSS as well, such as loss of certain transcripts due to lack of restriction enzyme recognition site and ambiguity in tag annotation.

The high sensitivity and absolute gene expression certainly favors MPSS. However, the technology is only available through Lynxgen Therapeutics, Inc. (then Solexa Inc till 2006 and then Illumina). Massively Parallel Signature Sequencing (MPSS) Introduction Massively Parallel Signature Sequencing (MPSS) is an open-ended platform that analyses the level of gene expression in a sample by counting the number of individual mRNA molecules produced by each gene. Tagged PCR products produced from cDNA are amplified so that each corresponding mRNA molecule gives ~100,000 of PCR products with a unique tag. Tags are used to attach the PCR products to microbeads. After several rounds of ligation-based sequence determination using the type IIs restriction endonuclease BbvI, a sequence signature of ~16-20 bp is identified from each bead; routinely, 17 bp of high quality sequence is obtained.

This is performed in parallel, and approximately 1,000,000 sequence signatures are obtained per experiment. Each signature sequence (MPSS tag) in a MPSS dataset is analyzed, compared with all other signatures and all identical signatures are counted. The level of expression of any single gene is calculated by dividing the number of signatures from that gene by the total number of signatures for all mRNAs present in the dataset. MPSS datasets are additive in nature, which means that datasets from multiple analyses with the same starting mRNA sample can be combined.

MPSS has routine sensitivity at a level of a few molecules of mRNA per cell, and the datasets are in a digital format that simplifies the management and analysis of the data.

How it Works

cDNA fragments are cloned onto micro-beads using the Lynx Megaclone technology. Starting with ome million mRNA molecules from a particular cell or tissue sample, Megaclone will produce one million beads, each containing 100,000 cloned copies of cDNA from each mRNA molecule. All molecules are covalently attached to the micro-beads at their poly(A) ends, so the DpnII end is available for the sequencing reactions.

Lynx Therapeutics' Massively Parallel Signature Sequencing (MPSS)

The first of the "next-generation" sequencing technologies, MPSS was developed in the 1990s at Lynx Therapeutics, a company founded in 1992 by Sydney Brenner and Sam Eletr. MPSS was a bead-based method that used a complex approach of adapter ligation followed by adapter decoding, reading the sequence in increments of four nucleotides; this method made it susceptible to sequence-specific bias or loss of specific sequences.

Because the technology was so complex, MPSS was only performed 'in-house' by Lynx Therapeutics and no machines were sold; when the merger with Solexa later led to the development of sequencing-by-synthesis, a more simple approach with numerous advantages, MPSS became obsolete.

However, the essential properties of the MPSS output were typical of later "next-gen" data types, including hundreds of thousands of short DNA sequences. In the case of MPSS, these were typically used for sequencing cDNA for measurements of gene expression levels. Lynx Therapeutics merged with Solexa in 2004, and this company was later purchased by Illumina.

Polony Sequencing

Polony sequencing, developed in the laboratory of George Church at Harvard, was among the first next-generation sequencing systems used to sequence a full genome in 2005. It combined an in vitro paired-tag library with emulsion PCR, an automated microscope, and ligation-based sequencing chemistry to sequence an E. coli genome at an accuracy of > 99.9999% and a cost approximately 1/10 that of Sanger sequencing. The technology was licensed to Agencourt Biosciences, subsequently spun out into Agencourt Personal Genomics, and ultimately incorporated into the Applied Biosystems SOLiD platform.

454 Pyrosequencing

454 Life Sciences, is a biotechnology company based in Branford, Connecticut. It is a subsidiary of Roche, and specializes in high-throughput DNA sequencing. A parallelized version of pyrosequencing was developed by 454 Life Sciences, which has since been acquired by Roche Diagnostics. The method amplifies DNA inside water droplets in an oil solution (emulsion PCR), with each droplet containing a single DNA template attached to a single primer-coated bead that then forms a clonal colony. The sequencing machine contains many picolitre-volume wells each containing a single bead and sequencing enzymes.

Pyrosequencing uses luciferase to generate light for detection of the individual nucleotides added to the nascent DNA, and the combined data are used to generate sequence read-outs. This technology provides intermediate read length and price per base compared to Sanger sequencing on one end and Solexa and SOLiD on the other. History and Major Achievements. Life Sciences was founded by Jonathan Rothberg originally as 454 Corporation, a subsidiary of CuraGen Corporation. For their method for low-cost gene sequencing, 454 Life Sciences were awarded the Wall Street Journal's Gold Medal for Innovation in the Biotech-Medical category in 2005. The name 454 was the code name by which the project was referred to at CuraGen and the numbers have no special meaning. In late March, 2007, Roche Diagnostics announced an agreement to purchase 454 Life Sciences for US$154.9 million. It will remain a separate business unit. In November 2006, Rothberg, Michael Egholm, and colleagues at 454 published a cover article with Svante Paabo in Nature describing the first million base pairs of the Neanderthal genome, and initiated the Neanderthal Genome Project to complete the sequence of the Neanderthal genome by 2009. In May 2007, Project "Jim", a project initiated by Rothberg and 454 Life Sciences to determine the first sequence of an individual was completed after sequencing the genome of James Dewey Watson, co-discoverer of the structure of DNA.

Technology

Sequencing uses a large-scale parallel pyrosequencing system capable of sequencing roughly 400-600 megabases of DNA per 10-hour run on the Genome Sequencer FLX with GS FLX Titanium series reagents. The system relies on fixing nebulized and adapter-ligated DNA fragments to small DNA-capture beads in a water-in-oil emulsion. The DNA fixed to these beads is then amplified by PCR. Each DNA-bound bead is placed into a ~29 ìm well on a PicoTiterPlate, a fiber optic chip. A mix of enzymes such as DNA polymerase, ATP sulfurylase, and luciferase are also packed into the well. The PicoTiterPlate is then placed into the GS FLX System for sequencing.

454 has experienced rapid growth since its acquisition by Roche Diagnostics and release of the GS20 sequencing machine in 2005, the first next-generation DNA sequencer on the market. In 2008, 454 Sequencing launched the GS FLX Titanium series reagents for use on the Genome Sequencer FLX instrument, with the ability to sequence 400-600 million base pairs per run with 400-500 base pair read lengths. The company has said it plans to launch kits enabling sequencing read lengths of up to 1,000 bp in 2010. In late 2009, 454 Life Sciences introduced the GS Junior System, a bench top version of the Genome Sequencer FLX System.

DNA Library Preparation and emPCR

Genomic DNA is fractionated into smaller fragments (300-800 base pairs) and polished (made blunt at each end). Short adaptors are then ligated onto the ends of the fragments. These adaptors provide priming sequences for both amplification and sequencing of the sample-library fragments. One adaptor (Adaptor B) contains a 5'-biotin tag for immobilization of the DNA library onto streptavidin-coated beads. After nick repair, the non-biotinylated strand is released and used as a single-stranded template DNA (sstDNA) library. The sstDNA library is assessed for its quality and the optimal amount (DNA copies per bead) needed for emPCR is determined by titration.

The sstDNA library is immobilized onto beads. The beads containing a library fragment carry a single sstDNA molecule. The bead-bound library is emulsified with the amplification reagents in a water-in-oil mixture. Each bead is captured within its own microreactor where PCR amplification occurs. This results in bead-immobilized, clonally amplified DNA fragments.

Sequencing

Single-stranded template DNA library beads are added to the DNA Bead Incubation Mix (containing DNA polymerase) and are layered with Enzyme Beads (containing sulfurylase and luciferase) onto a PicoTiterPlate device. The device is centrifuged to deposit the beads into the wells. The layer of Enzyme Beads ensures that the DNA beads remain positioned in the wells during the sequencing reaction. The bead-deposition process is designed to maximize the number of wells that contain a single amplified library bead.

The loaded PicoTiterPlate device is placed into the Genome Sequencer FLX Instrument. The fluidics sub-system delivers sequencing reagents (containing buffers and nucleotides) across the wells of the plate. The four DNA nucleotides are added sequentially in a fixed order across the PicoTiterPlate device during a sequencing run. During the nucleotide flow, millions of copies of DNA bound to each of the beads are sequenced in parallel. When a nucleotide complementary to the template strand is added into a well, the polymerase extends the existing DNA strand by adding nucleotide(s). Addition of one (or more) nucleotide(s) generates a light signal that is recorded by the CCD camera in the instrument. This technique is based on sequencing-by-synthesis and is called pyrosequencing.

The signal strength is proportional to the number of nucleotides; for example, homopolymer stretches, incorporated in a single nucleotide flow generate a greater signal than single nucleotides. However, the signal strength for homopolymer stretches is linear only up to eight consecutive nucleotides after which the signal falls-off rapidly. Data are stored in standard flowgram format (SFF) files for downstream analysis.

Applications

Sequencing can sequence any double-stranded DNA and enables a variety of applications including de novo whole genome sequencing, re-sequencing of whole genomes and target DNA regions, metagenomics and RNA analysis.

Full Genome Sequencing (de novo sequencing and resequencing)

Full genome sequencing (FGS), also referred to as whole genome sequencing (WGS), aims to sequence the entire genome of an organism, for example, humans, dogs, mice, viruses or bacteria. In June 2006, 454 Life Sciences launched a project with the Max Planck Institute for Evolutionary Anthropology to sequence the genome of the Neanderthal, the extinct closest relative of humans.. In September 2008 the complete Neanderthal mitochondrial genome was sequenced, establishing the divergence between humans and Neanderthal at 660,000 +/- 140,000 years, and the full genome was published in 2010, using a combination of 454 and Illumina sequencing.

Amplicon Sequencing

Amplicon (ultra deep) sequencing is a new field which is largely being enabled through 454 Sequencing technology. This method is designed to allow mutations to be detected at extremely low levels, and PCR amplify specific, targeted regions of DNA. This method is used to identify low frequency somatic mutations in cancer samples or discovery of rare variants in HIV infected individuals.

Transcriptome Sequencing

Transcriptome sequencing encompasses experiments including small RNA profiling and discovery, mRNA transcript expression analysis (full-length mRNA, expressed sequence tags (ESTs) and ditags, and allele-specific expression) and the sequencing and analysis of full-length mRNA transcripts. The transcriptome data derived from the Genome Sequencer FLX is ideally suited to detailed transcriptome investigation in the areas of novel gene discovery, gene space identification in novel genomes, assembly of full-length genes, single nucleotide polymorphism (SNP), insertion-deletion and splice-variant discovery.

Metagenomics

Metagenomics is the study of the genomic content in a complex sample. The two primary goals of this approach are to characterize the organisms present in a sample and identify what roles each organism has within a specific environment. Metagenomics samples are found nearly everywhere, including several microenvironments within the human body, soil samples, extreme environments such as deep mines and the various layers within the ocean. The Genome Sequencer FLX System enables a comprehensive view into the diversity of an environmental habitat. The system's long reads ensure the enormous specificity needed to compare sequenced reads against

DNA or protein databases. The platform is used to count environmental gene tags to analyze the relative abundance of microbial species under varying environmental conditions.

SOLiD Sequencing

Applied Biosystems' SOLiD technology employs sequencing by ligation. Here, a pool of all possible oligonucleotides of a fixed length are labeled according to the sequenced position. Oligonucleotides are annealed and ligated; the preferential ligation by DNA ligase for matching sequences results in a signal informative of the nucleotide at that position. Before sequencing, the DNA is amplified by emulsion PCR. The resulting bead, each containing only copies of the same DNA molecule, are deposited on a glass slide. The result is sequences of quantities and lengths comparable to Illumina sequencing. ABI Solid Sequencing

SOLiD (Sequencing by Oligonucleotide Ligation and Detection) is a next-generation sequencing technology developed by Life Technologies and has been commercially available since 2008. These next generation technologies generate hundreds of millions to billions of small sequence reads at one time. Well-known examples of such DNA sequencing methods include 454 pyrosequencing (introduced in 2005, generating millions of 200-400bp reads in 2009), the Solexa system (introduced in 2006, generating hundreds of millions of 50-100bp reads in 2009) and the SOLiD system (introduced in 2007, generating billions of 50bp reads in 2009). These methods have reduced the cost from $0.01/base in 2004 to nearly $0.0001/base in 2006 and increased the sequencing capacity from 1,000,000 bases/machine/day in 2004 to more than 5,000,000,000 bases/machine/day in 2009. Over 30 publications exist describing its use first for nucleosome positioning from Valouev et al., transcriptional profiling or strand sensitive RNA-Seq with Cloonan et al., single cell transcriptional profiling with Tang et al. and ultimately human resequencing with McKernan et al.

Chapter 4

Gene Technology

Genes generally express their functional effect through the production of proteins, which are complex molecules responsible for most functions in the cell. Proteins are made up of one or more polypeptide chains, each of which is composed of a sequence of amino acids, and the DNA sequence of a gene (through an RNA intermediate) is used to produce a specific amino acid sequence. This process begins with the production of an RNA molecule with a sequence matching the gene's DNA sequence, a process called transcription. This messenger RNA molecule is then used to produce a corresponding amino acid sequence through a process called translation. Each group of three nucleotides in the sequence, called a codon, corresponds either to one of the twenty possible amino acids in a protein or an instruction to end the amino acic sequence; this correspondence is called the genetic code. The flow of information is unidirectional: information is transferred from nucleotide sequences into the amino acid sequence of proteins, but it never transfers from protein back into the sequence of DNA—a phenomenon Francis Crick called the central dogma of molecular biology.

The specific sequence of amino acids results in a unique three-dimensional structure for that protein, and the three-dimensional structures of proteins are related to their functions. Some are simple structural molecules, like the fibers formed by the protein collagen. Proteins can bind to other proteins and simple molecules, sometimes acting as enzymes by facilitating chemical reactions within the bound molecules (without changing the structure of the protein itself). Protein structure is dynamic; the protein hemoglobin bends into slightly different forms as it facilitates the capture, transport, and release of oxygen molecules within mammalian blood.

A single nucleotide difference within DNA can cause a change in the amino acid sequence of a protein. Because protein structures are the result of their amino acid sequences, some changes can dramatically change the properties of a protein by destabilizing the structure or changing the surface of the protein in a way that changes its interaction with other proteins and molecules.

For example, sickle-cell anemia is a human genetic disease that results from a single base difference within the coding region for the â-globin section of hemoglobin, causing a single amino acid change that changes hemoglobin's physical properties. Sickle-cell versions of hemoglobin stick to themselves, stacking to form fibers that distort the shape of red blood cells carrying the protein. These sickle-shaped cells no longer flow smoothly through blood vessels, having a tendency to clog or degrade, causing the medical problems associated with this disease. Some genes are transcribed into RNA but are not translated into protein products—such RNA molecules are called non-coding RNA. In some cases, these products fold into structures which are involved in critical cell functions (e.g. ribosomal RNA and transfer RNA). RNA can also have regulatory effect through hybridization interactions with other RNA molecules (e.g. microRNA).

Nature Versus Nurture

Although genes contain all the information an organism uses to function, the environment plays an important role in determining the ultimate phenotype—a phenomenon often referred to as "nature vs. nurture". The phenotype of an organism depends on the interaction of genetics with the environment. One example of this is the case of temperature-sensitive mutations. Often, a single amino acid change within the sequence of a protein does not change its behavior and interactions with other molecules, but it does destabilize the structure. In a high temperature environment, where molecules are moving more quickly and hitting each other, this results in the protein losing its structure and failing to function. In a low temperature environment, however, the protein's structure is stable and it functions normally. This type of mutation is visible in the coat coloration of Siamese cats, where a mutation in an enzyme responsible for pigment production causes it to destabilize and lose function at high temperatures. The protein remains functional in areas of skin that are colder—legs, ears, tail, and face—and so the cat has dark fur at its extremities. Environment also plays a dramatic role in effects of the human genetic disease phenylketonuria. The mutation that causes phenylketonuria disrupts the ability of the body to break down the amino acid phenylalanine, causing a toxic build-up of an intermediate molecule that, in turn, causes severe symptoms of progressive mental retardation and seizures. If someone with the phenylketonuria mutation follows a strict diet that avoids this amino acid, however, they remain normal and healthy.

A popular method to determine how much role nature and nurture play is to study identical and fraternal twins or siblings of multiple birth. Because identical siblings come from the same zygote they are genetically the same. Fraternal siblings however are as different genetically from one another as normal siblings. By comparing how often the twin of a set has the same disorder between fraternal and identical twins, scientists can see whether there is more of a nature or nurture effect. One famous example of a multiple birth study includes the Genain quadruplets, who were identical quadruplets all diagnosed with schizophrenia.

Gene Regulation

The genome of a given organism contains thousands of genes, but not all these genes need to be active at any given moment. A gene is expressed when it is being transcribed into mRNA (and translated into protein), and there exist many cellular methods of controlling the expression of genes such that proteins are produced only when needed by the cell. Transcription factors are regulatory proteins that bind to the start of genes, either promoting or inhibiting the transcription of the gene.

Within the genome of Escherichia coli bacteria, for example, there exists a series of genes necessary for the synthesis of the amino acid tryptophan. However, when tryptophan is already available to the cell, these genes for tryptophan synthesis are no longer needed.

The presence of tryptophan directly affects the activity of the genes—tryptophan molecules bind to the tryptophan repressor (a transcription factor), changing the repressor's structure such that the repressor binds to the genes. The tryptophan repressor blocks the transcription and expression of the genes, thereby creating negative feedback regulation of the tryptophan synthesis process. Differences in gene expression are especially clear within multicellular organisms, where cells all contain the same genome but have very different structures and behaviors due to the expression of different sets of genes.

All the cells in a multicellular organism derive from a single cell, differentiating into variant cell types in response to external and intercellular signals and gradually establishing different patterns of gene expression to create different behaviors. As no single gene is responsible for the development of structures within multicellular organisms, these patterns arise from the complex interactions between many cells.

Within eukaryotes there exist structural features of chromatin that influence the transcription of genes, often in the form of modifications to DNA and chromatin that are stably inherited by daughter cells. These features are called "epigenetic" because they exist "on top" of the DNA sequence and retain inheritance from one cell generation to the next.

Because of epigenetic features, different cell types grown within the same medium can retain very different properties. Although epigenetic features are generally dynamic over the course of development, some, like the phenomenon of paramutation, have multigenerational inheritance and exist as rare exceptions to the general rule of DNA as the basis for inheritance.

Genetic Change

Mutations

During the process of DNA replication, errors occasionally occur in the polymerization of the second strand. These errors, called mutations, can have an impact on the phenotype of an organism, especially if they occur within the protein coding sequence of a gene. Error rates are usually very low—1 error in every 10–100 million bases—due to the "proofreading" ability of DNA polymerases. (Without proofreading error rates are a thousandfold higher; because many viruses rely on DNA and RNA polymerases that lack proofreading ability, they experience higher mutation rates.)

Processes that increase the rate of changes in DNA are called mutagenic: mutagenic chemicals promote errors in DNA replication, often by interfering with the structure of base-pairing, while UV radiation induces mutations by causing damage to the DNA structure. Chemical damage to DNA occurs naturally as well, and cells use DNA repair mechanisms to repair mismatches and breaks in DNA—nevertheless, the repair sometimes fails to return the DNA to its original sequence.

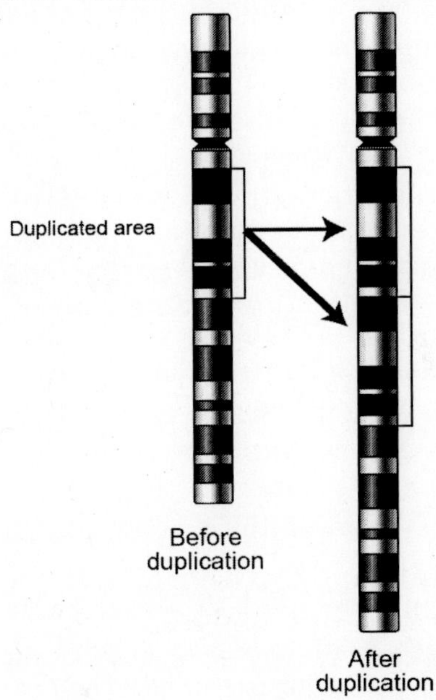

Figure : *Gene duplication allows diversification by providing redundancy: one gene can mutate and lose its original function without harming the organism.*

In organisms that use chromosomal crossover to exchange DNA and recombine genes, errors in alignment during meiosis can also cause mutations. Errors in crossover are especially likely when similar sequences cause partner chromosomes to adopt a mistaken alignment; this makes some regions in genomes more prone to mutating in this way. These errors create large structural changes in DNA sequence—duplications, inversions or deletions of entire regions, or the accidental exchanging of whole parts between different chromosomes (called translocation).

Natural Selection and Evolution

Natural Selection

Mutations alter an organisms genotype and occasionally this causes different phenotypes to appear. Most mutations have little effect on an organism's phenotype, health, or reproductive fitness. Mutations that do have an effect are usually deleterious, but occasionally some can be beneficial. Studies in the fly Drosophila melanogaster suggest that if a mutation changes a protein produced by a gene, about 70 percent of these mutations will be harmful with the remainder being either neutral or weakly beneficial.

Population genetics studies the distribution of genetic differences within populations and how these distributions change over time. Changes in the frequency of an allele in a population are mainly influenced by natural selection, where a given allele provides a selective or reproductive advantage to the organism, as well as other factors such as mutation, genetic drift, genetic draft, artificial selection and migration.

Over many generations, the genomes of organisms can change significantly, resulting in the phenomenon of evolution. Selection for beneficial mutations can cause a species to evolve into forms better able to survive in their environment, a process called adaptation. New species are formed through the process of speciation, often caused by geographical separations that prevent populations from exchanging genes with each other. The application of genetic principles to the study of population biology and evolution is referred to as the modern synthesis.

By comparing the homology between different species' genomes it is possible to calculate the evolutionary distance between them and when they may have diverged (called a molecular clock). Genetic comparisons are generally considered a more accurate method of characterizing the relatedness between species than the comparison of phenotypic characteristics.

The evolutionary distances between species can be used to form evolutionary trees; these trees represent the common descent and divergence of species over time, although they do not show the transfer of genetic material between unrelated species (known as horizontal gene transfer and most common in bacteria).

Research and Technology

Model Organisms

Although geneticists originally studied inheritance in a wide range of organisms, researchers began to specialize in studying the genetics of a particular subset of organisms. The fact that significant research already existed for a given organism would encourage new researchers to choose it for further study, and so eventually a few model organisms became the basis for most genetics research. Common research topics in model organism genetics include the study of gene regulation and the involvement of genes in development and cancer.

Organisms were chosen, in part, for convenience—short generation times and easy genetic manipulation made some organisms popular genetics research tools. Widely used model organisms include the gut bacterium Escherichia coli, the plant Arabidopsis thaliana, baker's yeast (Saccharomyces cerevisiae), the nematode Caenorhabditis elegans, the common fruit fly (Drosophila melanogaster), and the common house mouse (Mus musculus).

Medicine

Medical genetics seeks to understand how genetic variation relates to human health and disease. When searching for an unknown gene that may be involved in a disease, researchers commonly use genetic linkage and genetic pedigree charts to find the location on the genome associated with the disease. At the population level, researchers take advantage of Mendelian randomization to look for locations in the genome that are associated with diseases, a method especially useful for multigenic traits not clearly defined by a single gene. Once a candidate gene is found, further research is often done on the corresponding gene (called an orthologous gene) in model organisms. In addition to studying genetic diseases, the increased availability of genotyping methods has led to the field of pharmacogenetics—studying how genotype can affect drug responses.

Individuals differ in their inherited tendency to develop cancer, and cancer is a genetic disease. The process of cancer development in the body is a combination of events. Mutations occasionally occur within cells in the body as they divide. Although these mutations will not be inherited by any offspring, they can affect the behavior of cells, sometimes causing them to grow and divide more frequently. There are biological mechanisms that attempt to stop this process; signals are given to inappropriately dividing cells that should trigger cell death, but sometimes additional mutations occur that cause cells to ignore these messages. An internal process of natural selection occurs within the body and eventually mutations accumulate within cells to promote their own growth, creating a cancerous tumor that grows and invades various tissues of the body.

Normally, a cell divides only in response to signals: "growth factors", it stops growing when making contact with surrounding cells (contact inhibition), and in response to growth inhibitory signals, it divides a limited number of times and dies (apoptosis), it stays inside the epithelium and is not able to migrate to invade other organs.

To become a cancer cell, a cell has to accumulate mutations in a number of genes (3-7) that allow it to bypass all these regulations: it no longer needs growth factors to divide, it continues growing when making contact to neighbor cells, and ignores inhibitory signals, it will keep growing indefinitely and is immortal, it will escape from the epithelium and ultimately may be able to escape from the primary tumor, cross the endothelium of a blood vessel, be transported by the bloodstream and will colonize a new organ, forming deadly metastasis.

Although there are some genetic predispositions in a small fraction of cancers, the major fraction is due to a set of new genetic mutations that originally appear and accumulate in one or a small number of cells that will divide to form the tumor and are not transmitted to the progeny (somatic mutations).

The most frequent mutations are a loss of function of p53 protein, a tumor suppressor, or in the p53 pathway, and gain of function mutations in the ras proteins, or in other oncogenes.

For Human Genetic Diseases and Genetic Disorders

Research Methods

DNA can be manipulated in the laboratory. Restriction enzymes are commonly used enzymes that cut DNA at specific sequences, producing predictable fragments of DNA. DNA fragments can be visualized through use of gel electrophoresis, which separates fragments according to their length. The use of ligation enzymes allows DNA fragments to be connected, and by ligating fragments of DNA together from different sources, researchers can create recombinant DNA. Often associated with genetically modified organisms, recombinant DNA is commonly used in the context of plasmids—short circular DNA fragments with a few genes on them. By inserting plasmids into bacteria and growing those bacteria on plates of agar (to isolate clones of bacteria cells), researchers can clonally amplify the inserted fragment of DNA (a process known as molecular cloning). (Cloning can also refer to creating clonal organisms, by various means.)

DNA can also be amplified using a procedure called the polymerase chain reaction (PCR). By using specific short sequences of DNA, PCR can isolate and exponentially amplify a targeted region of DNA. Because it can amplify from extremely small amounts of DNA, PCR is also often used to detect the presence of specific DNA sequences.

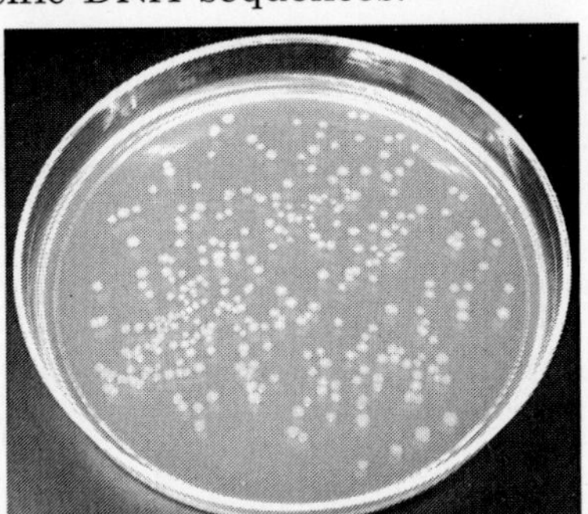

Figure : *Colonies of E. coli on a plate of agar, an example of cellular cloning and often used in molecular cloning.*

DNA Sequencing and Genomics

One of the most fundamental technologies developed to study genetics, DNA sequencing allows researchers to determine the sequence of nucleotides in DNA fragments. Developed in 1977 by Frederick Sanger and coworkers, chain-termination sequencing is now routinely used to sequence DNA fragments. With this technology, researchers have been able to study the molecular sequences associated with many human diseases. As sequencing has become less expensive, researchers have sequenced the genomes of many organisms, using computational tools to stitch together the sequences of many different fragments (a process called genome assembly).

These technologies were used to sequence the human genome, leading to the completion of the Human Genome Project in 2003. New high-throughput sequencing technologies are dramatically lowering the cost of DNA sequencing, with many researchers hoping to bring the cost of resequencing a human genome down to a thousand dollars. The large amount of sequence data available has created the field of genomics, research that uses computational tools to search for and analyze patterns in the full genomes of organisms. Genomics can also be considered a subfield of bioinformatics, which uses computational approaches to analyze large sets of biological data.

Quantitative Genetics

Quantitative genetics is the study of continuous traits (such as height or weight) and their underlying mechanisms. It is effectively an extension of simple Mendelian inheritance in that the combined effects of one or more genes and the environments in which they are expressed give rise to continuous distributions of phenotypic values.

History

The field was founded, in evolutionary terms, by the originators of the modern synthesis, R.A. Fisher, Sewall Wright and J. B. S. Haldane, and aimed to predict the response to selection given data on the phenotype and relationships of individuals. Analysis of Quantitative Trait Loci, or QTL, is a more recent addition to the study of quantitative genetics. A QTL is a region in the genome that affects the trait or traits of interest. Quantitative trait loci approaches require accurate phenotypic, pedigree and genotypic data from a large number of individuals.

Basic Principles

The phenotypic value (P) of an individual is the combined effect of the genotypic value (G) and the environmental deviation (E):

$$P = G + E$$

The genotypic value is the combined effect of all the genetic effects, including nuclear genes, mitochondrial genes and interactions between the genes. It is worthwhile to note that the mathematics is related to the genetics: for which the brief following revision may be useful. In disomic (diploid) organisms, a nucleus gene is represented twice in the gene-set ("genotype"), one contribution being provided by each parent during sexual reproduction. Each "gene" is located at a particular place (a "locus" - the Latin word for place; plural "loci") on corresponding chromosomes (homologues), one from each parent. Any gene may have several functional forms in the species as a whole, and each of these may lead to outwardly different "effects" (= an average result in the phenotype considered over a large sample of gene-backgrounds and environments).

These functional forms are "alleles" (or "allelomorphs", the original term). If both alleles at a gene have the same phenotypic effect (are the "same"), the gene is said to be "homozygous": if each allele at a gene is different in effect, the gene is "heterozygous". The average phenotypic outcome may also depend upon how alleles interact with their own homologous partner in the disomic genotype ("dominance"), and on how these alleles interact with those of other genes at other loci which also affect this phenotypic trait ("epistasis"). Notice that we have combined classical genetics ideas with those of statistics in this exposition.

Terms such as "gene", "homologue", "allele", "homozygous" and "heterozygous" are genetical: but "effect" is statistical, and refers to the average observed over an infinity of backgrounds, both genetical and environmental. Thus, we have very sneakily defined the "genetic value" (or "genotypic" value) as the infinity mean of all the phenotypes it can ever produce in time and space! Before molecular genetics, there simply was no other way to do it! And after molecular genetics, this is still the most utilitarian way to tackle the idea of genetic value! Also notice that we have openly used fundamentals of reproductive biology behind the genetics.

The founder of Quantitative Genetics - Sir Ronald Fisher - perceived all of this when he proposed the first mathematics of this branch of Genetics [Fisher R.A. (1930). The Genetical theory of Natural Selection. Oxford Clarendon Press.]. He sought to define a single statistical summary of all the variance arising from phenotypic change during the course of genetical assortment and segregation, which he called the "genetic" variance. His residual genotypic variance (which he called simply the "residual") represented that part of assortment which did not lead to phenotypic change, although the genes themselves had in fact been subject to meiosis and syngamy, of course. [A simplified exposition of this can be found in Falconer and Mackay (1966).] These partitions subsequently became the familiar subdivisions of the "additive" (A) and "dominance" (D) variances, respectively.

These later names are utterly misleading and very unfortunate, and have led to much confusion as to what they mean genetically. [It would have been better for posterity had they been named "Assortative" (A), and "Stable" (S).] A more gene-focused partitioning was invented by Mather and Jinks in 1971, but they also were statisticians rather than geneticists, had their own rather opaque symbolism, and became somewhat overwhelmed by the Fisherian approach. Added to all of this was the problem that Fisher's underlying reproduction model (mating system) was unrealistically simplistic: whilst it facilitated solving the equations, it didn't describe many real-life scenarios. Fortunately, Wright (in the 1950s) did provide the means to overcome complex mating systems, but its own complexity minimised its popular adoption.

Early in this millennium [Heredity (2003) 91: 85-89], a comprehensive linking together of all of these approaches reconciled their various meanings and relationships (as well as correcting an error), and suggested that new partitions reflecting real homozygosis [a] and heterozygosis [d] should replace the present "additive" [A] and "dominance" [D] subdivisions.

This paper revealed, by the way, that the Additive Genetical variance consists of all the homozygote variance, plus part of the dominance variance, and a frequency-weighted covariance between homozygote and heterozygote gene effects. The so-called Dominance variance contains only the remainder of the overall dominance variance of the gene in question, being therefore very misleading indeed!

It should be understood, however, that either method of partitioning still accounts for all of the genotypic variance in the model being used: it's the way it has been divided up which is being debated. At least the Environmental variance is much more straight-forward. This can be subdivided into a pure environmental component (E) and an interaction component (I) describing the gene-environment interaction. The overall "single gene" model can be written as:

$$P = a + d + E + I.$$

Expansion of the model to multiple genes (loci) is still not resolved satisfactorily, and until that is solved it is not possible to account for epistasis. The problem is being tackled currently. The contribution of those components cannot be determined in a single individual, but they can be estimated for whole populations by estimating the variances for those components, denoted as:

$$V_P = V_a + V_d + V_E + V_I$$

The heritability of a trait is the proportion of the total (i.e. phenotypic) variance (V_P) that is explained by the total genotypic variance (V_G). This is known also as the "broad sense" heritability (H). If only Additive genetic variance (V_A) is used in the numerator, the heritability is "narrow sense" (h). Unfortunately, this is often simply called "heritability", with little reflection about its true meaning. The broadsense heritability indicates the genotypic determination of the phenotype: while the latter estimates the degree of assortative disequlibrium in the trait.

Fisher proposed that this narrow-sense heritability might be appropriate in considering the results of natural selection, focusing as it does on disequilibrium: and it has been used also for predicting the results of artificial selection. This latter usage seems to be inappropriate, however, as breeders are interested in steering attributes towards new phenotypes (that is in utilising all the gene effects), rather than simply exploiting disequilibrium.

Resemblance Between Relatives

Central in estimating the variances for the various components is the principle of relatedness. A child has a father and a mother. Consequently, the child and father share 50% of their alleles, as do the child and the mother. However, the mother and father normally do not share alleles as a result of shared ancestors. Similarly, two full siblings share also on average 50% of the alleles with each other, while half siblings share only 25% of their alleles. This variation in relatedness can be used to estimate which proportion of the total phenotypic variance (V_P) is explained by the above-mentioned components.

Correlated Traits

Although some genes have only an effect on a single trait, many genes have an effect on various traits. Because of this, a change in a single gene will have an effect on all those traits. This is calculated using covariances, and the phenotypic covariance (Cov_P) between two traits can be partitioned in the same way as the variances described above. The genetic correlation is calculated by dividing the covariance between the additive genetic effects of two traits by the square root of the product of the variances for the additive genetic effects of the two traits:

$$Genetic\ correlation = \frac{Cov(A_1, A_2)}{\sqrt{V_{A1} * V_{A_2}}}$$

Genetic Distance

Genetic distance refers to the genetic divergence between species or between populations within a species. It is measured by a variety of parameters. Smaller genetic distances indicate a close genetic relationship whereas large genetic distances indicate a more distant genetic relationship. Genetic distance can be used to compare the genetic similarity between different species, such as humans and chimpanzees. Within a species genetic distance can be used to measure the divergence between different sub-species.

In its simplest form, the genetic distance between two populations is the difference in frequencies of a trait. For example the frequency of RH negative individuals is 50.4%

among Basques, 41.2% in France and 41.1% in England. Thus the genetic difference between the Basques and French is 9.2% and the genetic difference between the French and the English is 0.1% for the RH negative trait. The genetic distance of several individual traits can then be averaged to compute an overall genetic distance.

Measures of Genetic Distance

There are several measures used to indicate genetic distance. These include:

Fixation Index

A commonly used measure of genetic distance is the fixation index which varies between 0 and 1. A value of 0 indicates that two populations are genetically identical whereas a value of 1 indicates that two populations are different species.

Nei's Standard Genetic Distance

This measure assumes that genetic differences arise due to mutations and genetic drift.

Reynolds, Weir, and Cockerham's 1983

This measure assumes that genetic differences arise due to genetic drift only. A chromosome is an organized structure of DNA and protein found in cells. It is a single piece of coiled DNA containing many genes, regulatory elements and other nucleotide sequences. Chromosomes also contain DNA-bound proteins, which serve to package the DNA and control its functions. Chromosomes vary widely between different organisms.

The DNA molecule may be circular or linear, and can be composed of 100,000 to 10,000,000,000 nucleotides in a long chain. Typically, eukaryotic cells (cells with nuclei) have large linear chromosomes and prokaryotic cells (cells without defined nuclei) have smaller circular chromosomes, although there are many exceptions to this rule. Also, cells may contain more than one type of chromosome; for example, mitochondria in most eukaryotes and chloroplasts in plants have their own small chromosomes. In eukaryotes, nuclear chromosomes are packaged by proteins into a condensed structure called chromatin. This allows the very long DNA molecules to fit into the cell nucleus. The structure of chromosomes and chromatin varies through the cell cycle. Chromosomes are the essential unit for cellular division and must be replicated, divided, and passed successfully to their daughter cells so as to ensure the genetic diversity and survival of their progeny. Chromosomes may exist as either duplicated or unduplicated. Unduplicated chromosomes are single linear strands, whereas duplicated chromosomes contain two identical copies (called chromatids) joined by a centromere.

Compaction of the duplicated chromosomes during mitosis and meiosis results in the classic four-arm structure (pictured to the right). Chromosomal recombination plays a vital role in genetic diversity. If these structures are manipulated incorrectly, through processes known as chromosomal instability and translocation, the cell may undergo mitotic catastrophe and die, or it may unexpectedly evade apoptosis leading to the progression of cancer.

In practice "chromosome" is a rather loosely defined term. In prokaryotes and viruses, the term genophore is more appropriate when no chromatin is present. However, a large body of work uses the term chromosome regardless of chromatin content. In prokaryotes, DNA is usually arranged as a circle, which is tightly coiled in on itself, sometimes accompanied by one or more smaller, circular DNA molecules called plasmids. These small circular genomes are also found in mitochondria and chloroplasts, reflecting their bacterial origins. The simplest genophores are found in viruses: these DNA or RNA molecules are short linear or circular genophores that often lack structural proteins. The word chromosome comes from the Greek ÷ñöìá (chroma, colour) and óöìá (soma, body) due to their property of being very strongly stained by particular dyes.

History

In a series of experiments beginning in the mid-1880s, Theodor Boveri gave the definitive demonstration that chromosomes are the vectors of heredity. His two principles were the continuity of chromosomes and the individuality of chromosomes. It is the second of these principles that was so original. Wilhelm Roux suggested that each chromosome carries a different genetic load.

Boveri was able to test and confirm this hypothesis. Aided by the rediscovery at the start of the 1900s of Gregor Mendel's earlier work, Boveri was able to point out the connection between the rules of inheritance and the behaviour of the chromosomes.

Boveri influenced two generations of American cytologists: Edmund Beecher Wilson, Walter Sutton and Theophilus Painter were all influenced by Boveri (Wilson and Painter actually worked with him). In his famous textbook The Cell in Development and Heredity, Wilson linked together the independent work of Boveri and Sutton (both around 1902) by naming the chromosome theory of inheritance the "Sutton-Boveri Theory" (the names are sometimes reversed). Ernst Mayr remarks that the theory was hotly contested by some famous geneticists: William Bateson, Wilhelm Johannsen, Richard Goldschmidt and T.H. Morgan, all of a rather dogmatic turn of mind. Eventually, complete proof came from chromosome maps in Morgan's own lab.

In Eukaryotes

Eukaryotes (cells with nuclei such as those found in plants, yeast, and animals) possess multiple large linear chromosomes contained in the cell's nucleus. Each chromosome has one centromere, with one or two arms projecting from the centromere, although, under most circumstances, these arms are not visible as such. In addition, most eukaryotes have a small circular mitochondrial genome, and some eukaryotes may have additional small circular or linear cytoplasmic chromosomes. In the nuclear chromosomes of eukaryotes, the uncondensed DNA exists in a semi-ordered structure, where it is wrapped around histones (structural proteins), forming a composite material called chromatin.

Chromatin

Chromatin is the complex of DNA and protein found in the eukaryotic nucleus, which packages chromosomes. The structure of chromatin varies significantly between different stages of the cell cycle, according to the requirements of the DNA.

Interphase Chromatin

During interphase (the period of the cell cycle where the cell is not dividing), two types of chromatin can be distinguished:

- Euchromatin, which consists of DNA that is active, e.g., being expressed as protein.
- Heterochromatin, which consists of mostly inactive DNA. It seems to serve structural purposes during the chromosomal stages. Heterochromatin can be further distinguished into two types:
  - o Constitutive heterochromatin, which is never expressed. It is located around the centromere and usually contains repetitive sequences.
  - o Facultative heterochromatin, which is sometimes expressed.

Individual chromosomes cannot be distinguished at this stage – they appear in the nucleus as a homogeneous tangled mix of DNA and protein.

Metaphase Chromatin and Division

In the early stages of mitosis or meiosis (cell division), the chromatin strands become more and more condensed. They cease to function as accessible genetic material (transcription stops) and become a compact transportable form.

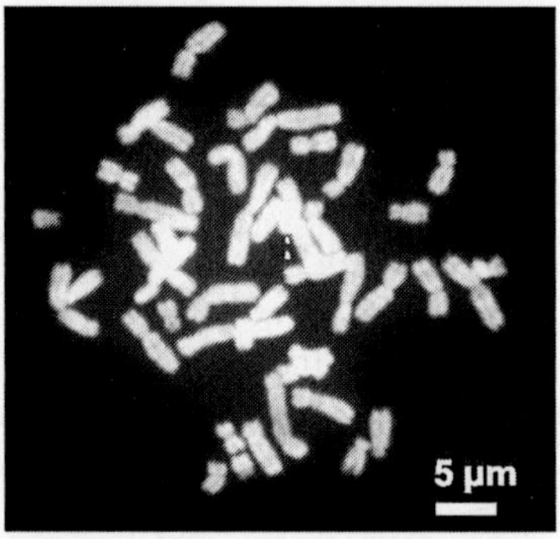

***Figure** : Human chromosomes during metaphase.*

This compact form makes the individual chromosomes visible, and they form the classic four arm structure, a pair of sister chromatids attached to each other at the centromere. The shorter arms are called p arms (from the French petit, small) and the longer arms are called q arms (q follows p in the Latin alphabet; q-g "grande"). This is the only natural context in which individual chromosomes are visible with an optical microscope.

During mitosis, microtubules grow from centrosomes located at opposite ends of the cell and also attach to the centromere at specialized structures called kinetochores, one of which is present on each sister chromatid. A special DNA base sequence in the region of the kinetochores provides, along with special proteins, longer-lasting attachment in this region.

The microtubules then pull the chromatids apart toward the centrosomes, so that each daughter cell inherits one set of chromatids. Once the cells have divided, the chromatids are uncoiled and DNA can again be transcribed. In spite of their appearance, chromosomes are structurally highly condensed, which enables these giant DNA structures to be contained within a cell nucleus.

In Prokaryotes

The prokaryotes – bacteria and archaea – typically have a single circular chromosome, but many variations exist. Most bacteria have a single circular chromosome that can range in size from only 160,000 base pairs in the endosymbiotic bacterium Candidatus Carsonella ruddii, to 12,200,000 base pairs in the soil-dwelling bacterium Sorangium cellulosum. Spirochaetes of the genus Borrelia are a notable exception to this arrangement, with bacteria such as Borrelia burgdorferi, the cause of Lyme disease, containing a single linear chromosome.

Structure in Sequences

Prokaryotic chromosomes have less sequence-based structure than eukaryotes. Bacteria typically have a single point (the origin of replication) from which replication starts, whereas some archaea contain multiple replication origins. The genes in prokaryotes are often organized in operons, and do not usually contain introns, unlike eukaryotes.

DNA Packaging

Prokaryotes do not possess nuclei. Instead, their DNA is organized into a structure called the nucleoid. The nucleoid is a distinct structure and occupies a defined region of the bacterial cell. This structure is, however, dynamic and is maintained and remodeled by the actions of a range of histone-like proteins, which associate with the bacterial chromosome. In archaea, the DNA in chromosomes is even more organized, with the DNA packaged within structures similar to eukaryotic nucleosomes.

Bacterial chromosomes tend to be tethered to the plasma membrane of the bacteria. In molecular biology application, this allows for its isolation from plasmid DNA by centrifugation of lysed bacteria and pelleting of the membranes (and the attached DNA). Prokaryotic chromosomes and plasmids are, like eukaryotic DNA, generally supercoiled. The DNA must first be released into its relaxed state for access for transcription, regulation, and replication.

Number of Chromosomes in Various Organisms

Eukaryotes

These tables give the total number of chromosomes (including sex chromosomes) in a cell nucleus. For example, human cells are diploid and have 22 different types of autosome, each present as two copies, and two sex chromosomes. This gives 46 chromosomes in total. Other organisms have more than two copies of their chromosomes, such as bread wheat, which is hexaploid and has six copies of seven different chromosomes – 42 chromosomes in total. Normal members of a particular eukaryotic species all have the same number of nuclear chromosomes. Other eukaryotic chromosomes, i.e., mitochondrial and plasmid-like small chromosomes, are much more variable in number, and there may be thousands of copies per cell.

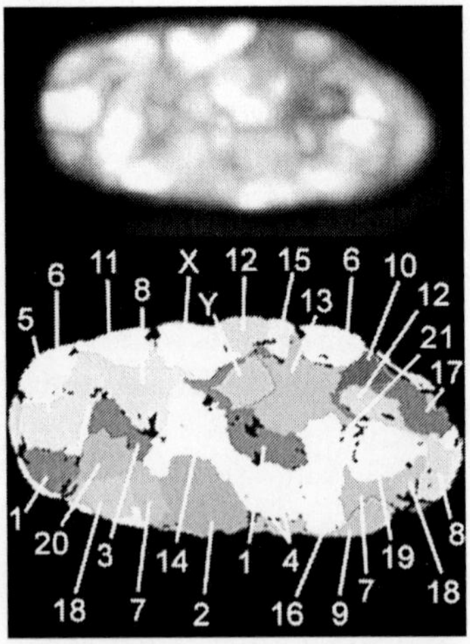

***Figure** : The 23 human chromosome territories during prometaphase in fibroblast cells.*

Asexually reproducing species have one set of chromosomes, which are the same in all body cells. However, asexual species can be either haploid or diploid. Sexually reproducing species have somatic cells (body cells), which are diploid [2n] having two sets of chromosomes, one from the mother and one from the father. Gametes, reproductive cells, are haploid [n]: They have one set of chromosomes. Gametes are produced by meiosis of a diploid germ line cell. During meiosis, the matching chromosomes of father and mother can exchange small parts of themselves (crossover), and thus create new chromosomes that are not inherited solely from either parent. When a male and a female gamete merge (fertilization), a new diploid organism is formed.

Some animal and plant species are polyploid [Xn]: They have more than two sets of homologous chromosomes. Plants important in agriculture such as tobacco or wheat are often polyploid, compared to their ancestral species. Wheat has a haploid number of seven chromosomes, still seen in some cultivars as well as the wild progenitors. The more-common pasta and bread wheats are polyploid, having 28 (tetraploid) and 42 (hexaploid) chromosomes, compared to the 14 (diploid) chromosomes in the wild wheat.

Prokaryotes

Prokaryote species generally have one copy of each major chromosome, but most cells can easily survive with multiple copies. For example, Buchnera, a symbiont of aphids has multiple copies of its chromosome, ranging from 10–400 copies per cell. However, in some large bacteria, such as Epulopiscium fishelsoni up to 100,000 copies of the chromosome can be present. Plasmids and plasmid-like small chromosomes are, as in eukaryotes, very variable in copy number. The number of plasmids in the cell is almost entirely determined by the rate of division of the plasmid – fast division causes high copy number.

Karyotype

In general, the karyotype is the characteristic chromosome complement of a eukaryote species. The preparation and study of karyotypes is part of cytogenetics.

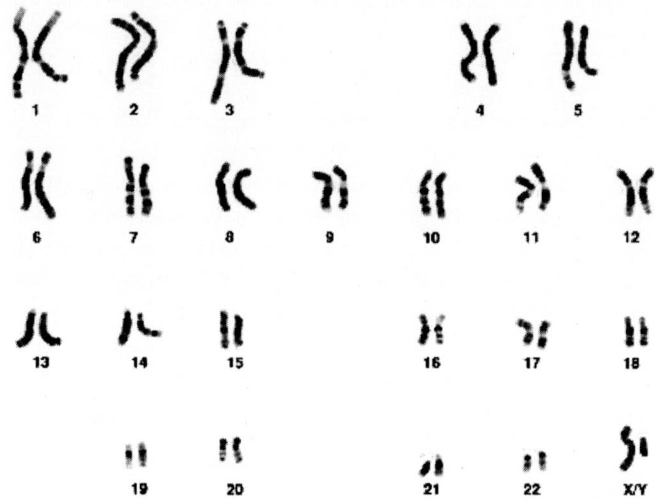

Figure : *Karyogram of a human male*

Although the replication and transcription of DNA is highly standardized in eukaryotes, the same cannot be said for their karyotypes, which are often highly variable. There may be variation between species in chromosome number and in detailed organization. In some cases, there is significant variation within species. Often there is:

1. variation between the two sexes
2. variation between the germ-line and soma (between gametes and the rest of the body)
3. variation between members of a population, due to balanced genetic polymorphism
4. geographical variation between races
5. mosaics or otherwise abnormal individuals.

Also, variation in karyotype may occur during development from the fertilised egg. The technique of determining the karyotype is usually called karyotyping. Cells can be locked part-way through division (in metaphase) in vitro (in a reaction vial) with colchicine. These cells are then stained, photographed, and arranged into a karyogram, with the set of chromosomes arranged, autosomes in order of length, and sex chromosomes (here X/Y) at the end: Like many sexually reproducing species, humans have special gonosomes (sex chromosomes, in contrast to autosomes). These are XX in females and XY in males.

Historical Note

Investigation into the human karyotype took many years to settle the most basic question: How many chromosomes does a normal diploid human cell contain? In 1912, Hans von Winiwarter reported 47 chromosomes in spermatogonia and 48 in oogonia, concluding an XX/XO sex determination mechanism. Painter in 1922 was not certain whether the diploid number of man is 46 or 48, at first favouring 46. He revised his opinion later from 46 to 48, and he correctly insisted on humans having an XX/XY system.

New techniques were needed to definitively solve the problem:

1. Using cells in culture
2. Pretreating cells in a hypotonic solution, which swells them and spreads the chromosomes
3. Arresting mitosis in metaphase by a solution of colchicine
4. Squashing the preparation on the slide forcing the chromosomes into a single plane
5. Cutting up a photomicrograph and arranging the result into an indisputable karyogram.

It took until 1954 before the human diploid number was confirmed as 46. Considering the techniques of Winiwarter and Painter, their results were quite remarkable. Chimpanzees (the closest living relatives to modern humans) have 48 chromosomes.

In Down Syndrome, there are three Copies of Chromosome 21

Chromosomal aberrations are disruptions in the normal chromosomal content of a cell and are a major cause of genetic conditions in humans, such as Down syndrome, although most aberrations have little to no effect. Some chromosome abnormalities do not cause disease in carriers, such as translocations, or chromosomal inversions, although they may lead to a higher chance of birthing a child with a chromosome disorder. Abnormal numbers of chromosomes or chromosome sets, called aneuploidy, may be lethal or may give rise to genetic disorders. Genetic counseling is offered for families that may carry a chromosome rearrangement.

The gain or loss of DNA from chromosomes can lead to a variety of genetic disorders. Human examples include:

- Cri du chat, which is caused by the deletion of part of the short arm of chromosome 5. "Cri du chat" means "cry of the cat" in French; the condition was so-named because affected babies make high-pitched cries that sound like those of a cat. Affected individuals have wide-set eyes, a small head and jaw, moderate to severe mental health issues, and are very short.
- Down syndrome, the most common trisomy, usually caused by an extra copy of chromosome 21 (trisomy 21). Characteristics include decreased muscle tone, stockier build, asymmetrical skull, slanting eyes and mild to moderate developmental disability.
- Edwards syndrome, or trisomy-18, the second-most-common trisomy. Symptoms include motor retardation, developmental disability and numerous congenital anomalies causing serious health problems. Ninety percent of those affected die in infancy; however, those that live past their first birthday are usually quite healthy thereafter. They have characteristic clenched hands and overlapping fingers.
- Isodicentric 15, also called idic(15), partial tetrasomy 15q, or inverted duplication 15 (inv dup 15).
- Jacobsen syndrome, which is very rare. It is also called the terminal 11q deletion disorder. Those affected have normal intelligence or mild developmental disability, with poor expressive language skills. Most have a bleeding disorder called Paris-Trousseau syndrome.
- Klinefelter's syndrome (XXY). Men with Klinefelter syndrome are usually sterile, and tend to be taller and have longer arms and legs than their peers. Boys with the syndrome are often shy and quiet, and have a higher incidence of speech delay and dyslexia. Without testosterone treatment, some may develop gynecomastia during puberty.

- Patau Syndrome, also called D-Syndrome or trisomy-13. Symptoms are somewhat similar to those of trisomy-18, without the characteristic folded hand.
- Small supernumerary marker chromosome. This means there is an extra, abnormal chromosome. Features depend on the origin of the extra genetic material. Cat-eye syndrome and isodicentric chromosome 15 syndrome (or Idic15) are both caused by a supernumerary marker chromosome, as is Pallister-Killian syndrome.
- Triple-X syndrome (XXX). XXX girls tend to be tall and thin and have a higher incidence of dyslexia.
- Turner syndrome (X instead of XX or XY). In Turner syndrome, female sexual characteristics are present but underdeveloped. Females with Turner syndrome often have a short stature, low hairline, abnormal eye features and bone development and a "caved-in" appearance to the chest.
- XYY syndrome. XYY boys are usually taller than their siblings. Like XXY boys and XXX girls, they are more likely to have learning difficulties.
- Wolf-Hirschhorn syndrome, which is caused by partial deletion of the short arm of chromosome 4. It is characterized by severe growth retardation and severe to profound mental health issues.

Chapter 5

Recombinant DNA Technology

Genetic engineering, also called genetic modification, is the direct human manipulation of an organism's genetic material in a way that does not occur under natural conditions. It involves the use of recombinant DNA techniques, but does not include traditional animal and plant breeding or mutagenesis. Any organism that is generated using these techniques is considered to be a genetically modified organism. The first organisms genetically engineered were bacteria in 1973 and then mice in 1974. Insulin producing bacteria were commercialized in 1982 and genetically modified food has been sold since 1994.

The most common form of genetic engineering involves the insertion of new genetic material at an unspecified location in the host genome. This is accomplished by isolating and copying the genetic material of interest, generating a construct containing all the genetic elements for correct expression, and then inserting this construct into the host organism. Other forms of genetic engineering include gene targeting and knocking out specific genes via engineered nucleases such as zinc finger nucleases or engineered homing endonucleases.

Genetic engineering techniques have been applied in numerous fields including research, biotechnology, and medicine. Medicines such as insulin and human growth hormone are now produced in bacteria, experimental mice such as the oncomouse and the knockout mouse are being used for research purposes and insect resistant and/or herbicide tolerant crops have been commercialized. Genetically engineered plants and animals capable of producing biotechnology drugs more cheaply than current methods (called pharming) are also being developed and in 2009 the FDA approved the sale of the pharmaceutical protein antithrombin produced in the milk of genetically engineered goats.

Definition

Genetic engineering alters the genetic makeup of an organism using techniques that introduce heritable material prepared outside the organism either directly into the host or into a cell that is then fused or hybridized with the host. This involves using recombinant nucleic acid (DNA or RNA) techniques to form new combinations of heritable genetic material followed by the incorporation of that material either indirectly through a vector system or directly through micro-injection, macro-injection and micro-encapsulation techniques.

Genetic engineering does not include traditional animal and plant breeding, in vitro fertilisation, induction of polyploidy, mutagenesis and cell fusion techniques that do not use recombinant nucleic acids or a genetically modified organism in the process. Cloning and stem cell research, although not considered genetic engineering, are closely related and genetic engineering can be used within them. Synthetic biology is an emerging discipline that takes genetic engineering a step further by introducing artificially synthesized genetic material from raw materials into an organism.

If genetic material from another species is added to the host, the resulting organism is called transgenic. If genetic material from the same species or a species that can naturally breed with the host is used the resulting organism is called cisgenic. Genetic engineering can also be used to remove genetic material from the target organism, creating a knock out organism. In Europe genetic modification is synonymous with genetic engineering while within the United States of America it can also refer to conventional breeding methods.

History

Humans have altered the genomes of species for thousands of years through artificial selection and more recently mutagenesis. Genetic engineering as the direct manipulation of DNA by humans outside breeding and mutations has only existed since the 1970s. The term "genetic engineering" was first coined by Jack Williamson in his science fiction novel *Dragon's Island*, published in 1951, one year before DNA's role in heredity was confirmed by Alfred Hershey and Martha Chase, and two years before James Watson and Francis Crick showed that the DNA molecule has a double-helix structure.

In 1972 Paul Berg created the first recombinant DNA molecules by combined DNA from the monkey virus SV40 with that of the lambda virus. In 1973 Herbert Boyer and Stanley Cohen created the first transgenic organism by inserting antibiotic resistance genes into the plasmid of an *E. coli* bacterium. A year later Rudolf Jaenisch created a transgenic mouse by introducing foreign DNA into its embryo, making it the world's first transgenic animal. In 1976 Genentech, the first genetic engineering company was founded by Herbert Boyer and Robert Swanson and a year later and the company produced a human protein (somatostatin) in *E.coli*. Genentech announced the production of genetically engineered human insulin in 1978.

In 1980, the U.S. Supreme Court in the Diamond v. Chakrabarty case ruled that genetically altered life could be patented. The insulin produced by bacteria, branded humulin, was approved for release by the Food and Drug Administration in 1982. The first field trials of genetically engineered plants occurred in France and the USA in 1986, tobacco plants were engineered to be resistant to herbicides. The People's Republic of China was the first country to commercialize transgenic plants, introducing a virus-resistant tobacco in 1992. In 1994 Calgene attained approval to commercially release the Flavr Savr tomato, a tomato engineered to have a longer shelf life. In 1994, the European Union approved tobacco engineered to be resistant to the herbicide bromoxynil, making it the first genetically engineered crop commercialized in Europe.

In 1995, Bt Potato was approved safe by the Environmental Protection Agency, making it the first pesticide producing crop to be approved in the USA. In 2009 11 transgenic crops were grown commercially in 25 countries, the largest of which by area grown were the USA, Brazil, Argentina, India, Canada, China, Paraguay and South Africa. In 2010, scientists at the J. Craig Venter Institute, announced that they had created the first synthetic bacterial genome, and added it to a cell containing no DNA. The resulting bacterium, named Synthia, was the world's first synthetic life form.

Process

Isolating the Gene

First, the gene to be inserted into the genetically modified organism must be chosen and isolated. Presently, most genes transferred into plants provide protection against insects or tolerance to herbicides. In animals the majority of genes used are growth hormone genes. Once chosen the genes must be isolated. This typically involves multiplying the gene using polymerase chain reaction (PCR). If the chosen gene or the donor organism's genome has been well studied it may be present in a genetic library. If the DNA sequence is known, but no copies of the gene are available, it can be artificially synthesized. Once isolated, the gene is inserted into a bacterial plasmid.

Constructs

The gene to be inserted into the genetically modified organism must be combined with other genetic elements in order for it to work properly. The gene can also be modified at this stage for better expression or effectiveness. As well as the gene to be inserted most constructs contain a promoter and terminator region as well as a selectable marker gene. The promoter region initiates transcription of the gene and can be used to control the location and level of gene expression, while the terminator region ends transcription. The selectable marker, which in most cases confers antibiotic resistance to the organism it is expressed in, is needed to determine which cells are transformed with the new gene. The constructs are made using recombinant DNA techniques, such as restriction digests, ligations and molecular cloning.

Gene Targeting

The most common form of genetic engineering involves inserting new genetic material randomly within the host genome. Other techniques allow new genetic material to be inserted at a specific location in the host genome or generate mutations at desired genomic loci capable of knocking out endogenous genes. The technique of gene targeting uses homologous recombination to target desired changes to a specific endogenous gene. This tends to occur at a relatively low frequency in plants and animals and generally requires the use of selectable markers. The frequency of gene targeting can be greatly enhanced with the use of engineered nucleases such as zinc finger nucleases, engineered homing endonucleases, or nucleases created from TAL effectors. In addition to enhancing gene targeting, engineered nucleases can also be used to introduce mutations at endogenous genes that generate a gene knockout.

Transformation

About 1% of bacteria are naturally able to take up foreign DNA but it can also be induced in other bacteria. Stressing the bacteria for example, with a heat shock or an electric shock, can make the cell membrane permeable to DNA that may then incorporate into their genome or exist as extrachromosomal DNA. DNA is generally inserted into animal cells using microinjection, where it can be injected through the cells nuclear envelope directly into the nucleus or through the use of viral vectors. In plants the DNA is generally inserted using *Agrobacterium*-mediated recombination or biolistics.

In *Agrobacterium*-mediated recombination the plasmid construct must also contain T-DNA. *Agrobacterium* naturally inserts DNA from a tumour inducing plasmid into any susceptible plant's genome it infects, causing crown gall disease. The T-DNA region of this plasmid is responsible for insertion of the DNA. The genes to be inserted are cloned into a binary vector, which contains T-DNA and can be grown in both *E. Coli* and *Agrobacterium*. Once the binary vector is constructed the plasmid is transformed into *Agrobacterium* containing no plasmids and plant cells are infected. The *Agrobacterium* will then naturally insert the genetic material into the plant cells.

In biolistics particles of gold or tungsten are coated with DNA and then shot into young plant cells or plant embryos. Some genetic material will enter the cells and transform them. This method can be used on plants that are not susceptible to *Agrobacterium* infection and also allows transformation of plant plastids. Another transformation method for plant and animal cells is electroporation. Electroporation involves subjecting the plant or animal cell to an electric shock, which can make the cell membrane permeable to plasmid DNA. In some cases the electroporated cells will incorporate the DNA into their genome. Due to the damage caused to the cells and DNA the transformation efficiency of biolistics and electroporation is lower than agrobacterial mediated transformation and microinjection.

Selection

Not all the organism's cells will be transformed with the new genetic material; in most cases a selectable marker is used to differentiate transformed from untransformed cells. If a cell has been successfully transformed with the DNA it will also contain the marker gene. By growing the cells in the presence of an antibiotic or chemical that selects or marks the cells expressing that gene it is possible to separate the transgenic events from the non-transgenic. Another method of screening involves using a DNA probe that will only stick to the inserted gene. A number of strategies have been developed that can remove the selectable marker from the mature transgenic plant.

Regeneration

As often only a single cell is transformed with genetic material the organism must be regrown from that single cell. As bacteria consist of a single cell and reproduce clonally regeneration is not necessary. In plants this is accomplished through the use of tissue culture. Each plant species has different requirements for successful regeneration through tissue culture. If successful an adult plant is produced that contains the transgene in every cell. In animals it is necessary to ensure that the inserted DNA is present in the embryonic stem cells. When the offspring is produced they can be screened for the presence of the gene. All offspring from the first generation will be heterozygous for the inserted gene and must be mated together to produce a homozygous animal.

Confirmation

Further tests using PCR, Southern Blots and Bioassays are needed to confirm that the gene is expressed and functions correctly. The organism's offspring are also tested to ensure that the trait can be inherited and that it follows a Mendelian inheritance pattern.

Applications

Genetic engineering has applications in medicine, research, industry and agriculture and can be used on a wide range of plants, animals and micro organism.

Medicine

In medicine genetic engineering has been used to mass-produce insulin, human growth hormones, follistim (for treating infertility), human albumin, monoclonal antibodies, antihemophilic factors, vaccines and many other drugs. Vaccination generally involves injecting weak live, killed or inactivated forms of viruses or their toxins into the person being immunized. Genetically engineered viruses are being developed that can still confer immunity, but lack the infectious sequences. Mouse hybridomas, cells fused together to create monoclonal antibodies, have been humanised through genetic engineering to create human monoclonal antibodies.

Genetic engineering is used to create animal models of human diseases. Genetically modified mice are the most common genetically engineered animal model. They have been used to study and model cancer (the oncomouse), obesity, heart disease, diabetes, arthritis, substance abuse, anxiety, aging and Parkinson disease. Potential cures can be tested against these mouse models. Also genetically modified pigs have been bred with the aim of increasing the success of pig to human organ transplantation. Gene therapy is the genetic engineering of humans by replacing defective human genes with functional copies. This can occur in somatic tissue or germline tissue. If the gene is inserted into the germline tissue it can be passed down to that person's descendants.

Gene therapy has been used to treat patients suffering from immune deficiencies (notably Severe combined immunodeficiency) and trials have been carried out on other genetic disorders. The success of gene therapy so far has been limited and a patient (Jesse Gelsinger) has died during a clinical trial testing a new treatment. There are also ethical concerns should the technology be used not just for treatment, but for enhancement, modification or alteration of a human beings' appearance, adaptability, intelligence, character or behavior. The distinction between cure and enhancement can also be difficult to establish. Transhumanists consider the enhancement of humans desirable.

Research

Genetic engineering is an important tool for natural scientists. Genes and other genetic information from a wide range of organisms are transformed into bacteria for storage and modification, creating genetically modified bacteria in the process. Bacteria are cheap, easy to grow, clonal, multiply quickly, relatively easy to transform and can be stored at -80°C almost indefinitely. Once a gene is isolated it can be stored inside the bacteria providing an unlimited supply for research.

Organisms are genetically engineered to discover the functions of certain genes. This could be the effect on the phenotype of the organism, where the gene is expressed or what other genes it interacts with. These experiments generally involve loss of function, gain of function, tracking and expression.

- Loss of function experiments, such as in a gene knockout experiment, in which an organism is engineered to lack the activity of one or more genes. A knockout experiment involves the creation and manipulation of a DNA construct *in vitro*, which, in a simple knockout, consists of a copy of the desired gene, which has been altered such that it is non-functional. Embryonic stem cells incorporate the altered gene, which replaces the already present functional copy. These stem cells are injected into blastocysts, which are implanted into surrogate mothers. This allows the experimenter to analyze the defects caused by this mutation and thereby determine the role of particular genes. It is used especially frequently in developmental biology. Another method, useful in organisms such as Drosophila (fruit fly), is to induce mutations in a large population and then screen the progeny for the desired mutation. A similar process can be used in both plants and prokaryotes.
- Gain of function experiments, the logical counterpart of knockouts. These are sometimes performed in conjunction with knockout experiments to more finely establish the function of the desired gene. The process is much the same as that in knockout engineering, except that the construct is designed to increase the function of the gene, usually by providing extra copies of the gene or inducing synthesis of the protein more frequently.

- Tracking experiments, which seek to gain information about the localization and interaction of the desired protein. One way to do this is to replace the wild-type gene with a 'fusion' gene, which is a juxtaposition of the wild-type gene with a reporting element such as green fluorescent protein (GFP) that will allow easy visualization of the products of the genetic modification. While this is a useful technique, the manipulation can destroy the function of the gene, creating secondary effects and possibly calling into question the results of the experiment. More sophisticated techniques are now in development that can track protein products without mitigating their function, such as the addition of small sequences that will serve as binding motifs to monoclonal antibodies.
- Expression studies aim to discover where and when specific proteins are produced. In these experiments, the DNA sequence before the DNA that codes for a protein, known as a gene's promoter, is reintroduced into an organism with the protein coding region replaced by a reporter gene such as GFP or an enzyme that catalyzes the production of a dye. Thus the time and place where a particular protein is produced can be observed. Expression studies can be taken a step further by altering the promoter to find which pieces are crucial for the proper expression of the gene and are actually bound by transcription factor proteins; this process is known as promoter bashing.

Industrial

By engineering genes into bacterial plasmids it is possible to create a biological factory that can produce proteins and enzymes. Some genes do not work well in bacteria, so yeast, a eukaryote, can also be used. Bacteria and yeast factories have been used to produce medicines such as insulin, human growth hormone, and vaccines, supplements such as tryptophan, aid in the production of food (chymosin in cheese making) and fuels. Other applications involving genetically engineered bacteria being investigated involve making the bacteria perform tasks outside their natural cycle, such as cleaning up oil spills, carbon and other toxic waste.

Agriculture

One of the best-known and controversial applications of genetic engineering is the creation of genetically modified food. There are three generations of genetically modified crops. First generation crops have been commercialized and most provide protection from insects and/or resistance to herbicides. There are also fungal and virus resistant crops developed or in development. They have been developed to make the insect and weed management of crops easier and can indirectly increase crop yield.

The second generation of genetically modified crops being developed aim to directly improve yield by improving salt, cold or drought tolerance and to increase the nutritional value of the crops. The third generation consists of pharmaceutical crops, crops that contain edible vaccines and other drugs. Some agriculturally important animals have been genetically modified with growth hormones to increase their size while others have been engineered to express drugs and other proteins in their milk. The genetic engineering of agricultural crops can increase the growth rates and resistance to different diseases caused by pathogens and parasites. This is beneficial as it can greatly increase the production of food sources with the usage of fewer resources that would be required to host the world's growing populations. These modified crops would also reduce the usage of chemicals, such as fertilizers and pesticides, and therefore decrease the severity and frequency of the damages produced by these chemical pollution.

Ethical and safety concerns have been raised around the use of genetically modified food. A major safety concern relates to the human health implications of eating genetically modified food, in particular whether toxic or allergic reactions could occur. Gene flow into related non-transgenic crops, off target effects on beneficial organisms and the impact on biodiversity are important environmental issues. Ethical concerns involve religious issues, corporate control of the food supply, intellectual property rights and the level of labelling needed on genetically modified products.

Other Uses

In materials science, a genetically modified virus has been used to construct a more environmentally friendly lithium-ion battery. Some bacteria have been genetically engineered to create black and white photographs while others have potential to be used as sensors by expressing a fluorescent protein under certain environmental conditions. Genetic engineering is also being used to create BioArt and novelty items such as blue roses, and glowing fish.

Negative Genetic Engineering (Cures and Treatments)

When treating problems that arise from genetic disorder, one solution is gene therapy, also known as negative genetic engineering. A genetic disorder is a condition caused by the genetic code of the individual, such as spina bifida or autism. When this happens, genes may be expressed in unfavorable ways or not at all, and this generally leads to further complications. The idea of gene therapy is that a non-pathogenic virus or other delivery systems can be used to insert into DNA—a good copy of the gene—into cells of the living individual. The modified cells would divide as normal and each division would produce cells that express the desired trait. The result would be that he/she would then have the ability to express the trait that was previously absent, at least partially. This form of genetic engineering could help alleviate many problems, such as diabetes, cystic fibrosis, or other genetic diseases.

Positive Genetic Engineering (Enhancement)

The potential of genetic engineering to cure medical conditions opens the question of exactly what such a condition is. Some view aging and death as medical conditions and therefore potential targets for engineering solutions. They see human genetic engineering potentially as a key tool in this. The difference between cure and enhancement from this perspective is merely one of degree. Theoretically genetic engineering could be used to drastically change people's genomes, which could enable people to regrow limbs and other organs, perhaps even extremely complex ones such as the spine.

It could also be used to make people smarter, stronger, or to increase the capacity of the lungs, among other things. If a gene exists in nature, perhaps it could be changed into a human cell. In this view, there is no qualitative difference (only a quantitative one) between, for instance, a genetic intervention to cure muscular atrophy, and a genetic intervention to improve muscle function even when those muscles are functioning at or below the human average (since there is also an average muscle function for those with a particular type of dystrophy, which the treatment would improve upon).

Others feel, there is an important distinction between using genetic technologies to treat those who are suffering, and to make those who are already healthy seem more superior to the average person. Though theory and speculation suggest, that genetic engineering could be used to make people stronger, faster, smarter, or to increase lung capacity. The AAAS report finds that there is little evidence to support this theory. Can this currently be done without very unsafe and therefore unethical human experiments. Because different cells have different tasks, changing one cell to do a function differently, will not only affect that one task, but it can affect many other tasks as well.

Controversy

Ethics: The genetic engineering of humans has raised many controversial ethical issues. While negative genetic engineering (gene therapy) does indeed raise a debate, the use of genetic engineering for human enhancement arouses the strongest feelings on both sides. Genetic engineering is tested on animals, often including primates. Some animal rights activists find this inhumane. Genetic engineering can be used to cure peoples with diabetes. It is possible to extract genes from cells which are called beta cells and then to insert the insulin producing genes into a bacterium. Then the bacterium will start producing insulin. Genetic modification of embryos can pose an ethical question about the rights of the baby. One belief is that every fetus should be free to not be genetically modified. Others believe that parents hold the rights to change their unborn children. Still others believe that every child should have the right to be born free from preventable diseases.

Molecular Biologist Lee M. Silver believes that unlike Aldous Huxley's Brave New World, where a totalitarian government controls all of the genetic enhancements (they actually use eugenics instead of direct genetic modification) in society, the use of gene therapy to design children will be spread through what he calls "free market eugenics" (Silver 315). Wealthy families will opt to design their child with genetic advantages because other families are doing so, and everybody wants to provide their newborn child with the best opportunities in life, with a leg up on the competition.

The greatest fear for Silver is that we will design so many children with germline gene therapy, that the families wealthy enough to design their children, will pass down these enhanced traits to future generations. This gene therapy will obviously cost money, and the less wealthy families will be left to procreate naturally, and introduce their children into the world disadvantaged from their first breath.

The impact on society will be a new alignment of classes, no longer will we separate people by their ethnic differences, the new division will be between what Silver calls 'the naturals' and 'the GenRich', or genetically enhanced. The major worry here is that the 'genetic gulf' between these two classes will become so wide that humans will become separate species (Silver 313).

In Popular Culture

- *Maximum Ride series by James Patterson:* The main characters are six human children who had bird DNA injected into them while they were in their mothers' wombs.
- *Mobile Suit Gundam SEED (anime):* Set in a world in which genetically modified humans, termed 'Coordinators', have been ostracized and isolated from unmodified humans, termed 'Naturals'. Due to extreme differences in mental and physical abilities between the two groups, racial, economic, and political issues have arisen, culminating in war. Gundam Seed addresses such concerns as animosity caused by the jealousy of Naturals over Coordinator abilities, both groups looking down on one another as being lesser life forms, and genocidal factions emerging on both sides. The series primarily explores these issues from the point of view of a Coordinator protagonist who finds himself fighting on the side of the Naturals, as his childhood friend has become a member of the Coordinator military, giving a perspective on both sides of the conflict.

- *Gattaca (film):* Presents a biopunk vision of a society driven by new eugenics. Children of the middle and upper classes are selected through preimplantation genetic diagnosis to ensure they possess the best hereditary traits of their parents.
- *Oryx and Crake by Margaret Atwood:* Apocalyptic pseudo-dystopian story, one of the chief plot points of which involves the genetic engineering of a new type of transhuman and the pathological decimation of *Homo sapiens*.
- *Bioshock (Video Game):* The main enemies the player encounters over the course of the game are known as splicers, so called for their genetic manipulation, or gene splicing. In the game a compound called ADAM is responsible for the genetic manipulation. The compound is harvested from a species of sea slugs, it acts like a seemingly benign form of cancer, destroying native cells and replacing them with the unstable stem versions. The splicers have all become so addicted to the ADAM that they will kill anything or anyone for it.
- *Old Man's War series by John Scalzi:* To create an army of soldiers capable of defending the human race from endless alien hordes the Colonial Defense Forces recruits 75 year olds and gives them new younger bodies capable of super human feats.
- *Batman Beyond:* In the episode *Splicers* and making appearances later in the series, a new fad of splicing animal genes for cosmetic and enhancement purposes makes the teen scene. When research shows that splicing increases aggression in users, resulting it being banned in Gotham, only finding a place in the criminal underworld later. Notable Splicers are Woof of the Jokerz (spliced with hyena DNA), and the Zander (spliced with T-Rex DNA), leader of KOBRA.
- *Halo (series) (Video Game):* the protagonist and several other characters in the Halo universe known as Spartans have all gone through comprehensive genetic augmentation, making them almost super human. The augmentations vary from Muscular Enhancement Injections to a Catalytic Thyroid Implant. The effects

include nearly unbreakable bones, increased reflexes and increased muscle tissue with increased density. The SPARTAN-III program also included a mutagen that increased aggression and injury tolerance.

- *Deus Ex:* The protagonists of both games in the series are genetically modified by injections called nano-augs (Deus Ex) and bio mods (Invisible War). These nanite injections alter the host's genes to enhance them with new skill that would help them through different obstacles in the game. Both the protagonists and antagonists are bio-modified along with other support characters and neutral characters.
- *PROTOTYPE:* The story is centered around a genetic engineering company called Gentek. The engineers in Gentek modified a chimeran-virus to make it 10 times as deadly, to 'unlock' previously dormant parts of a subjects DNA, and they made it to the point of where it could copy those infected with it down to the genetic level. The new 'infected' being then has complete and total control over its genetic structure, allowing it to develop instantaneous shape-shifting abilities. However, the game reveals that the only 'infected' being that can do this is a person already dead (or dying) when the virus enters the bloodstream. The virus must also be in a state of emergency to survive and not die out.
- *Warhammer 40,000:* In the games, and table-top series, Space marines are the result of Genetic Engineering, by replacing organs, and limbs that have been engineered. The Marine himself, also goes through rigorous physical and mental training, to make him accustom to the newly added appendages and organs.

GM Food Controversy

The genetically modified foods controversy is a dispute over the relative advantages and disadvantages of genetically modified (GM) food crops and other uses of genetically-modified organisms in food production. The dispute involves biotechnology companies, governmental regulators, non-governmental organizations and scientists. The dispute is most intense in Japan and Europe, where public concern about GM food is higher than in other parts of the world such as the United States. In the United States GM crops are more widely grown and the introduction of these products has been less controversial. The five key areas of political controversy related to genetically engineered food are food safety, the effect on natural ecosystems, gene flow into non GE crops, moral/religious concerns, and corporate control of the food supply. To date, not a single instance of harm to human health or the environment has been documented with GM crops. Several benefits have been widely accepted and are uncontested in the scientific literature. These include reductions in insecticide use on GE cotton, enhanced biological diversity in GE cotton fields (compared to non-GE fields), enhanced farmer income and communal benefits, increased yields for poor farmers and improved health of farmworkers. Although the use of herbicide tolerant crops remain controversial, because of the need to spray herbicides, it is clear that the use of these crops has promoted a shift to less toxic herbicides.

The concept of substantial equivalence has been established to address the fears of some consumers about the safety of GE crops. "Substantial equivalence embodies the concept that if a new food or food component is found to be substantially equivalent to an existing food or food component, it can be treated in the same manner with respect to safety (i.e., the food or food component can be concluded to be as safe as the conventional food or food component)" (Joint FAO/WHO Biotechnology and Food Safety Report, 1996). The rationale for this approach is that it would be impossible to test all the new crop varieties every year for food safety. Only a few food crops on the market have been shown to cause adverse health effects and all of these were the result of conventional genetic modification, not from genetic engineering To date no adverse health effects caused by products approved for sale have been documented, although two products failed initial safety testing and were discontinued, due to allergic reactions.

Most feeding trials have observed no toxic effects and saw that GM foods were equivalent in nutrition to unmodified foods, although a few non-peer-reviewed reports speculate physiological changes to GM food. Although there is now broad scientific consensus that GE crops on the market are safe to eat, some scientists and advocacy groups such as Greenpeace and World Wildlife Fund call for additional and more rigorous testing before marketing genetically engineered food. Another area of controversy is what effect pest and herbicide-resistant crops have on ecosystems. For example, whereas 14 years of studies indicate cotton engineered for resistance to insects has resulted in massive reduction in insecticides globally, the use of crops engineered for herbicide tolerance has changed herbicide patterns, with a clear shift to use of less toxic herbicides. Studies assessing diversity of non-target insects has revealed enhanced biological diversity in GE fields compared to non-GE fields. The risk and effects of have also been cited as concerns, with the possibility that genes might spread from modified crops to wild relatives. There is no evidence that the risk of pollen transfer is greater for any of the GE crops on the market as compared to non-GE crops.

Health Risks and Benefits

Present Knowledge on GM Food Safety

Worldwide, there is a range of perspectives within non-governmental organizations on the safety of GM foods. For example, the US pro-GM group AgBioWorld has argued that GM foods have been proven safe, while other pressure groups and consumer rights groups, such as the Organic Consumers Association, and Greenpeace claim the long-term health risks which GM could pose, or the environmental risks associated with GM, have not yet been adequately investigated. In Japan, Consumers Union of Japan are opposed to GMO foods. They also claim that truly independent research in these areas is systematically blocked by the GM corporations which own the GM seeds and reference materials.

A 2008 review published by the Royal Society of Medicine noted that GM foods have been eaten by millions of people worldwide for over 15 years, with no reports of ill effects. Similarly a 2004 report from the US National Academies of Sciences stated: "To date, no adverse health effects attributed to genetic engineering have been documented in the human population." A 2004 review of feeding trials in the *Italian Journal of Animal Science* found no differences among animals eating genetically modified plants.

A 2005 review in *Archives of Animal Nutrition* concluded that first-generation genetically modified foods had been found to be similar in nutrition and safety to non-GM foods, but noted that second-generation foods with "significant changes in constituents" would be more difficult to test, and would require further animal studies. However, a 2009 review in *Nutrition Reviews* found that although most studies concluded that GM foods do not differ in nutrition or cause any detectable toxic effects in animals, some studies did report adverse changes at a cellular level caused by some GM foods, concluding that "More scientific effort and investigation is needed to ensure that consumption of GM foods is not likely to provoke any form of health problem". A review published in 2009 by Dona and Arvanitoyannis concluded that "results of most studies with GM foods indicate that they may cause some common toxic effects such as hepatic, pancreatic, renal, or reproductive effects and may alter the hematological, biochemical, and immunologic parameters".

However responses to this review in 2009 and 2010 note that the Dona and Arvanitoyannis concentrated on articles with an anti-GM bias that have been refuted by scientists in peer-reviewed articles elsewhere - for example the 35S promoter, stability of transgenes, antibiotic marker genes and the claims for toxic effects of GM foods. In 2007, a review by Domingo of the toxicity by searching in the Publimed data base using 12 search terms, cited 68 references, found that the "number of references" on the safety of GM/transgenic crops was "surprisingly limited" and questioned whether the safety of genetically modified food has been demonstrated; the review also remarked that its conclusions were in agreement with three earlier reviews by Zdunczyk (2001), Bakshi (2003), and Pryme and Lembcke (2003). However, an article in 2007 by Vain found 692 research studies focusing on GM crop and food safety and identified a strong increase in the publication of such articles in recent years.

Vain commented that the multidisciplinarian nature of GM research complicates the retrieval of GM studies and requires using many search terms (he used more than 300) and multiple data bases.

Safety Assessments

The starting point for the safety assessment of genetically engineered food products is to assess if the food is "substantially equivalent" to its natural counterpart. To decide if a modified product is substantially equivalent, the product is tested by the manufacturer for unexpected changes in a limited set of components such as toxins, nutrients or allergens that are present in the unmodified food.

If these tests show no significant difference between the modified and unmodified products, then no further food safety testing is required. The manufacturers data is then assessed by an independent regulatory body, such as the U.S. Food and Drug Administration. However, if the product has no natural equivalent, or shows significant differences from the unmodified food, then further safety testing is carried out.

A 2003 review in *Trends in Biotechnology* identified 7 main parts of a standard safety test:

1. Study of the introduced DNA and the new proteins or metabolites that it produces;
2. Analysis of the chemical composition of the relevant plant parts, measuring nutrients, anti-nutrients as well as any natural toxins or known allergens;
3. Assess the risk of gene transfer from the food to microorganisms in the human gut;
4. Study the possibility that any new components in the food might be allergens;
5. Estimate how much of a normal diet the food will make up;
6. Estimate any toxicological or nutritional problems revealed by this data;
7. Additional animal toxicity tests if there is the possibility that the food might pose a risk.

This process was examined further in a review published by Kuiper *et al.* 2002 in the journal *Toxicology*, which stated that substantial equivalence does not itself measure risks, but instead identifies differences between existing products and new foods, which might pose dangers to health. If differences do exist, identifying these differences is a starting point for a full safety assessment, rather than an end point. The authors concluded that "The concept of substantial equivalence is an adequate tool in order to identify safety issues related to genetically modified products that have a traditional counterpart".

However, the review also noted difficulties in applying this standard in practice, including the fact that traditional foods contain many chemicals that have toxic or carcinogenic effects and that our existing diets therefore have not been proven to be safe. This lack of knowledge on unmodified food poses a problem, as GM foods may have differences in anti-nutrients and natural toxins that have never been identified in the original plant, raising the possibility that harmful changes could be missed. The application of substantial equivalence has also been more strongly criticized. For example, in a speech in 1999, Andrew Chesson of the University of Aberdeen, stated that substantial equivalence testing "could be flawed in some cases" and that some current safety tests could allow harmful substances to enter the human food chain.

In a commentary in *Nature* Millstone *et al.* argued that all GM foods should have extensive biological, toxicological and immunological tests and that the concept of substantial equivalence based solely on chemical analyzes of the components of a food should be abandoned. They stated that this is necessary since it is currently impossible to predict the biological properties of a substance only from knowledge of its chemistry. This commentary was controversial and was criticized for misleading presentation of data and presenting an over-simplified version of safety assessments. For example, Kuiper *et al.* responded to this criticism by noting that equivalency testing does involve more than chemical tests and may include toxicity testing.

Medical writer Barbara Keeler and Marc Lappé argued in a 2001 article in the Los Angeles Times that the differences between Genetically Modified and conventional foods challenges the presumption of equivalence. Using Roundup ready soy that has been on the market since 1995 as an example, they noted the differences when compared to its unmodified counterpart.

Significantly lower levels of protein than unmodified soy. Significantly lower levels of phenylalanine, an essential amino acid and as a dietary supplement, the reason doctors advise the consumption of soy products. Levels of trypsin inhibitor were 27% higher and after toasting, Lectin was double that found in conventional soy, both are known allergens. GM soy also has 29% less choline, a B-complex vitamin. Round up ready soy had also stunted the growth of rats in Monsanto's study but had not effected cattle although it had increased the fat content of their milk. The authors do not maintain that modified soy is a hazard but that the FDA accepting such significant differences as being *substantially equivalent* illustrates the need for more rigorous testing, and preferably not by the biotech industries themselves.

The value of current independent studies is problematic as, due to restrictive End-user agreements, researchers are forbidden by law from publishing independent research in peer reviewed journals without the approval of the agritech companies. Cornell University's Elson Shields, the spokesperson for a group of scientists who oppose this practice, submitted a statement to the United States Environmental Protection Agency (EPA) protesting that "as a result of restrictive access, no truly independent research can be legally conducted on many critical questions regarding the technology". Scientific American noted that several studies that were initially approved by seed companies were later blocked from publication when they returned "unflattering" results. While recognising that seed companies' intellectual property rights need to be protected, Scientific American calls the practice dangerous and has called for the restrictions on research in the End-user agreements to be lifted immediately and for the EPA to require, as a condition of approval, that independent researchers have unfettered access to GM products for testing.

The Welsh pressure group GM Free Cymru argues that governments should use independent studies rather than industry studies to assess crop safety. GM Free Cymru has also stated that independently funded researcher, Professor Bela Darvas of Debrecen University was refused Mon 863 Bt corn to use in his studies after previously publishing that a different variety of Monsanto corn was lethal to two Hungarian protected insect species and an insect classified as a rare.

Allergenicity

Worldwide, reports of allergies to all kinds of foods, particularly nuts, fish and shellfish, seem to be increasing, but it is not known if this reflects a genuine change in the risk of allergy, or an increased awareness of food allergies by the public. Some environmental organizations, such as the European Green Party and Greenpeace, have suggested that GM food might trigger food allergies, although other environmentalists have implicated causes as diverse as the greenhouse effect increasing pollen levels, greater exposure to synthetic chemicals, cleaner lifestyles, or more mold in buildings. A 2005 review in the journal *Allergy* of the results from allergen testing of current GM foods stated that "no biotech proteins in foods have been documented to cause allergic reactions".

A well-known case of a GM plant that did not reach the market due to it producing an allergic reaction was a new form of soybean intended for animal feed. The allergen was transferred unintentionally from the Brazil nut into genetically engineered soybeans, in a bid to improve soybean nutritional quality for animal feed use. This new protein increased the levels in the GM soybean of the natural essential amino acid methionine, which is commonly added to poultry feed. Investigation of the GM soybeans revealed that they produced immune reactions in people with Brazil nut allergies, since the methionine rich protein chosen by Pioneer Hi-Bred happened to be a major source of Brazil nut allergy.

Although this soybean strain was not developed as a human food, Pioneer Hi-Bred discontinued further development of the GM soybean, due to the difficulty in ensuring that none of these soybeans entered the human food chain.

In November 2005 a pest-resistant field pea developed by the Australian CSIRO for use as a pasture crop was shown to cause an allergic reaction in mice. Work on this variety was immediately halted. The protein added to the pea did not cause the reaction in humans or mice in isolation, but when it was expressed in the pea, it exhibited a subtly different structure which may have caused the allergic reaction. The immunologist who tested the pea noted that crops need to be evaluated case-by-case, and a Greenpeace activist commented that to his knowledge, no countries required feeding studies for approval of genetically modified foods.

Plant scientist Maarten J Chrispeels has made these comments about this example: The recent Prescott et al. paper in JFAC contains a very interesting study on the immunogenicity of amylase [starch digestion enzyme] inhibitor in its native form (isolated from beans) and expressed as a transgene in peas. First of all, amylase inhibitor is a food protein, but also a "toxic" protein because it inhibits our digestive amylases. This is one of the reasons you have to cook your beans! (The other toxic bean protein is phytohemagglutinin and it is much more toxic).

This particular amylase inhibitor is found in the common bean (other species have other amylase inhibitors). Even though it is a food protein, it is unlikely ever to be used for genetic engineering of human foods because it inhibits our amylases. What the results show is that the protein, when synthesized in pea cotyledons has a different immunogenicity than when it is isolated from bean cotyledons (the native form). This is somewhat surprising but may be related to the presence of slightly different carbohydrate chains.

These cases of products that failed safety testing can either be viewed as evidence that genetic modification can produce unexpected and dangerous changes in foods, or alternatively that the current tests are effective at identifying any safety problems before foods come on the market. Genetic modification can also be used to remove allergens from foods, which may, for example, allow the production of soy products that would pose a smaller risk of food allergies than standard soybeans.

A hypo-allergenic strain of soybean was tested in 2003 and shown to lack the major allergen that is found in the beans. A similar approach has been tried in ryegrass, which produces pollen that is a major cause of hayfever: here a fertile GM grass was produced that lacked the main pollen allergen, demonstrating that the production of hypoallergenic grass is also possible.

Environmental Risks and Benefits

The large scale growth of GM plants may have both positive and negative effects on the environment. These may be both direct effects, on organisms that feed on or interact with the crops, or wider effects on food chains produced by increases or decreases in the numbers of other organisms. As an example of benefits, insect-resistant Bt-expressing crops will reduce the number of pest insects feeding on these plants, but as there are fewer pests, farmers do not have to apply as much insecticide, which in turn tends to increase the number of non-pest insects in these fields. A 2006 study of the global impact of GM crops, published by the UK consultancy PG Economics and funded by Illinois-Missouri Biotechnology Alliance, concluded that globally, the technology reduced pesticide spraying by 286,000 tons in 2006, decreasing the environmental impact of herbicides and pesticides by 15%.

By reducing the amount of ploughing needed, GM technology led to reductions of greenhouse gases from soil equivalent to removing 6.56 million cars from the roads. However, a 2009 study published by the Organic Center stated that the use of genetically engineered corn, soybean, and cotton increased the use of herbicides by 383 million pounds (191,500 tons), and pesticide use by 318.4 million pounds (159,200 tons). As an example of a concern about environmental risk, a lab at Cornell University published an article which caused worry in the US that Bt-corn pollen might affect the monarch butterfly. However this concern was disproved by six comprehensive articles in the Proceedings of the National Academy of Sciences in 2001. Monarch populations increased, despite increased Bt corn probably due to reduced pesticide use. Other possible effects might come from the spread of genes from modified plants to unmodified relatives, which might produce species of weeds resistant to herbicides. In some areas of the US "superweeds" have evolved naturally, these weeds are resistant to herbicides and have forced farmers to return to traditional crop management practices.

There has been controversy over the results of a farm-scale trial in the United Kingdom comparing the impact of GM crops and conventional crops on farmland biodiversity. Some claimed that the results showed that GM crops had a significant negative impact on wildlife. They pointed out that the studies showed that using herbicide resistant GM crops allowed better weed control and that under such conditions there were fewer weeds and fewer weed seeds.

This result was then extrapolated to suggest that GM crops would have significant impact on the wildlife that might rely on farm weeds. The President of the Royal Society, the body that had carried out the trials, stated that "To generalize and declare 'all GM is bad' or 'all GM is good' for the environment as a result of these experiments is a gross over-simplification", arguing that although the trials showed that the combination of some GM crops with long-lasting herbicides were bad for biodiversity, using other GM crops without these herbicides increased biodiversity. In July 2005 British scientists showed that transfer of a herbicide-resistance gene from GM oilseed rape to a wild cousin, charlock, and wild turnips was possible. Many agricultural scientists and food policy specialists view GM crops as an important element in sustainable food security and environmental management. This point of view is summarized in the ABIC Manifesto: On our planet, 18% of the land mass is used for agricultural production. This fraction cannot be increased substantially.

It is absolutely essential that the yield per unit of land increases beyond current levels given that: The human population is still growing, and will reach about nine billion by 2040;70,000 km^2 of agricultural land (equivalent to 60% of the German agricultural area) are lost annually to growth of cities and other non-agricultural uses; Consumer diets in developing countries are increasingly changing from plant-based proteins to animal protein, a trend that requires a greater amount of crop-based feeds. Other scientists, such as Dr. Charles Benbrook, argue that improvement of global food security is hardly being addressed by genetic research and that a lack of yield is often not caused by insufficient genetic resources. Regarding the issues of intellectual property and patent law, an international report from the year 2000 states: If the rights to these tools are strongly and universally enforced - and not extensively licensed or provided pro bono in the developing world - then the potential applications of GM technologies described previously are unlikely to benefit the less developed nations of the world for a long time (ie until after the restrictions conveyed by these rights have expired).

Issues with Bt Maize

A well publicized claim associated with Bt crops (or transgenic maize) was the concern that pollen from Bt maize might kill the monarch butterfly. This report was puzzling because the pollen from most maize hybrids contains much lower levels of

Bt than the rest of the plant and led to multiple follow-up studies. One possible issue revealed in these studies is the possibility that the initial study was flawed; based on the way the pollen was collected, in that they collected and fed non-toxic pollen that was mixed with anther walls that did contain Bt toxin. A collaborative research exercise was carried out over two years by several groups of scientists in the US and Canada, looking at the effects of Bt pollen in both the field and the laboratory. This resulted in a risk assessment that concluded that any risk posed by the corn to butterfly populations under real-world conditions was negligible. The USDA has stated that the weight of the evidence is that Bt crops do not pose a risk to the monarch butterfly.

An independent 2002 review of the scientific literature concluded that "the commercial large-scale cultivation of current Bt–maize hybrids did not pose a significant risk to the monarch population" and noted that despite large-scale planting of GM crops, the butterfly's population is increasing.

In 2007 Andreas Lang, Éva Lauber and Béla Darvas criticized these studies, arguing that there can be a great difference in the effects between the acute exposure tested for and chronic exposure. Moreover, they stated that the "worst case conditions" performed were not in fact worst case scenarios, as laboratory conditions with ample food supply and a favorable climate ensure healthy subjects. They instead believe that in the wild, low temperatures, rain and parasites and disease might exacerbate a Bt effect on butterfly larvae. Their own experiments suggested that some butterfly species were negatively affected by such chronic exposure. Jörg Romeis, who conducted the original studies, replied that if species of Butterfly are affected as Darvas claims that a "more comprehensive assessment will be needed and, depending on the degree and nature of concern, this may extend to field testing".

A 2001 report in *Nature* presented evidence that Bt maize was cross-breeding with unmodified maize in Mexico, although the data in this paper was later described as originating from an artifact and *Nature* stated that "the evidence available is not sufficient to justify the publication of the original paper". A subsequent large-scale study, in 2005, failed to find any evidence of contamination in Oaxaca. However, other authors have stated that they also found evidence of cross-breeding between natural maize and transgenic maize. There is also a risk that for example, transgenic maize will crossbreed with wild grass variants, and that the Bt-gene will end up in a natural environment, retaining its toxicity. An event like this would have ecological implications. However, there is no evidence of crossbreeding between maize and wild grasses.

In 2009 it was reported that 82,000 hectares (200,000 acres) of Bt corn in South Africa failed to produce seeds. Monsanto claimed average yield was reduced by 25% in those fields affected, it compensated the farmers concerned and the corn varieties were affected by a mistake made in the seed breeding process. Marian Mayet an environmental activitist and director of the Africa-centre for biosecurity in Johannesburg called for a government investigation and asserted that the biotechnology was at fault, "*You cannot make a 'mistake' with three different varieties of corn*". In 2009 South African farmers planted 1,900,000 hectares (4,700,000 acres) of GM maize (73% of the total crop).

As of 2007, a phenomenon called Colony Collapse Disorder (CCD) was noticed in bee hives all over North America, and elsewhere. Although it is not certain if this is a new phenomenon, initial ideas on the possible causes ranged from poor nutrition, infections, parasites and pesticide use. More unusual speculations included radio waves from cellphone base stations, climate change, and the use of transgenic crops containing Bt. The Mid-Atlantic Apiculture Research and Extension Consortium published a report on 2007-03-27 that found no evidence that pollen from Bt crops is adversely affecting bees. Several researchers in the US have since attributed CCD to the spread of a new virus called Israeli acute paralysis virus, although other parasites have also been implicated.

Legal Issues in the US

Alfalfa

On 21 June 2010, the US Supreme Court in *Monsanto v. Geertson Seed Farms* issued its first ruling about GM crops. This case relates to Roundup Ready alfalfa and dates back to 2006, when a coalition of groups led by the Center for Food Safety raised concerns about environmental impacts that the

United States Department of Agriculture (USDA) allegedly failed to address before approving the planting of Roundup Ready alfalfa. In response, in 2007, Judge Charles Breyer of the California Northern District Court ruled that USDA was in error when it approved the planting of Roundup Ready alfalfa. In June 2009, a divided three-judge panel on the 9th U.S. Circuit Court of Appeals upheld Judge Charles Breyer's decision, and Monsanto appealed to the Supreme Court. The impact of the current US Supreme Court ruling is somewhat unclear, with both sides appearing to claim victory.

Sugar Beets

There is also a somewhat similar case involving sugar beets before the same California Northern District Court. This is a case involving Monsanto's breed of pesticide-resistant sugar beets. While Judge Jeffrey S. White (issuing his ruling in the spring of 2010) allowed the planting of GM sugar beets to continue, he also warned that this may be blocked in the future while an environmental review is taking place. Finally, on 13 August 2010, Judge Jeffrey S. White ordered the halt to the planting of the genetically modified sugar beets in the US. He indicated that "the Agriculture Department had not adequately assessed the environmental consequences before approving them for commercial cultivation." In 2010, before the ruling, 95% of the sugar beet grown in the US was GM. About half the sugar supply in the US came from sugar beet.

Control of the Market

Patent holder companies use their control of their own GMO to corner the market and gain profit. However, other companies often compete for the little market share available to GM foods worldwide. Detractors such as Greenpeace say that patent rights give corporations a dangerous amount of control over their product while corporations claim that they need product control in order to prevent seed piracy, fulfill financial obligations to shareholders and invest in further GM development. Governments have also sought to protect their commercial interests through punitive measures against countries resisting GM foods on ethical or scientific grounds: after moves in France to ban a Monsanto GM corn variety, the US embassy recommended that 'we calibrate a target retaliation list that causes some pain across the EU'.

Controversial Cases

Pusztai Potato Controversy

A large media event occurred in 1998 when scientist Árpád Pusztai, who was considered by some to be a world expert on plant lectins, reported in a television interview that he had found that rats fed potatoes genetically modified by the English biotech company *Cambridge Agricultural Genetics* to contain lectin, a natural insecticide in snowdrop plants, had stunted growth and immune system damage. Initially the Rowett Institute's director, Philip James, congratulated Pusztai after the interview, but his attitude changed after confusion arose over whether the lectin was from a jackbean or a snowdrop (the jackbean lectin is poisonous to the intestines).

A press release from the Rowett Institute stated that the lectin had come from a jackbean, but no experiments were made using lectins from the jackbean; in the confusion, it was alleged that Pusztai hadn't actually carried out the experiments. James suspended Pusztai and used misconduct procedures to seize the data. His annual contract was not renewed and both Pusztai and his wife were banned from speaking publicly.

In October 1998 the Rowett Institute published an audit criticizing Pusztai's results, which, along with Pusztai's raw data, was sent to six anonymous reviewers who criticized Pusztai's work. Pusztai responded that the raw data was "never intended for publication under intense scrutiny". Pusztai sent the audit report and his rebuttal to scientists who requested it, and in February 1999, twenty-one European and American scientists, mostly friends or acquaintances of Pusztai, orchestrated by the environmental group Friends of the Earth, released a memo supporting Pusztai. Stanley Ewen, who worked with Pusztai, conducted a followup study supporting Pusztai's work and presented the work to a lectin meeting in Sweden. In October 1999 Pusztai's research was published in the journal The Lancet.

Because of the controversial nature of his research the data in this paper, co-authored by Stanley Ewen, was seen by a total of six reviewers when presented for peer review; five of these reviewers judged the work acceptable, although one of the five "deemed the study flawed but favored publication to avoid suspicions of a conspiracy against Pusztai and to give colleagues a chance to see the data for themselves".

The paper did not mention stunted growth or immunity issues, but reported that rats fed on potatoes genetically modified with the snowdrop lectin had "thickening in the mucosal lining of their colon and their jejunum" when compared with rats fed on non modified potatoes. Three Dutch scientists criticized the study on the grounds that the unmodified potatoes were not a fair control diet, and that any rats fed only on potatoes will suffer from a protein deficiency; Pusztai responded to these criticisms by stating that the protein and energy were comparable, and that "a sample size of six is perfectly normal in studies like this". The genetically modified potatoes were later destroyed by *Cambridge Agricultural Genetics* as were all details of their modification.

Bioequivalence Study of a Corn Cultivar

A controversy arose around biotech company Monsanto's data on a 90-Day Rat Feeding Study on the MON863 strain of GM corn. In May 2005, critics of GM foods pointed to differences in kidney size and blood composition found in this study, suggesting that the observed differences raises questions about the regulatory concept of substantial equivalence. Anti-GM campaigner Jeffrey M.

Smith, writing in Biophile Magazine, quoting comments from Pusztai and Seralini,has stated that nutritional studies typically use young, fast-growing animals with starting weights not varying by more than 2% from the average whereas Monsanto's research design used a mix of young and old animals with starting weights ranging from 198.4 to 259.8 grams. Seralini and two other authors published a study of these data, funded by Greenpeace, in 2007 making similar points. The raising of this issue prompted the European Food Safety Authority (EFSA) to reexamine the safety data on this strain of corn. The EFSA concluded that the observed small numerical decrease in rat kidney weights were not biologically meaningful, and the weights were well within the normal range of kidney weights for control animals. There were no corresponding microscopic findings in the relevant organ systems, and they stated that all blood chemistry and organ weight values fell within the "normal range of historical control values" for rats. In addition the EFSA review stated that the statistical methods used by Séralini et al. in the analysis of the data were incorrect. The European Commission has approved the 863 corn for animal and human consumption. Food Standards Australia New Zealand reviewed the 2007 Seralini et al study and concluded that "...all of the statistical differences between rats fed MON 863 corn and control rats are attributable to normal biological variation."

Greenpeace stated in a 2007 press release that Séralini et al. had completed a similar analysis of the NK603 strain of corn and came to similar conclusions as they did in their previous study. Séralini et al included this in a re-analysis of three existing rat feeding studies published in 2009. The European Food Safety Authority reviewed the 2009 Seralini et al paper and concluded that the author's claims were not supported by the data in their paper. They noted that many of their fundamental statistical criticisms of the 2007 paper also applied to the 2009 paper. There was no new information that would change the EFSA's conclusions that the three GM maize types were safe for human, animal health and the environment The French High Council of Biotechnologies Scientific Committee (HCB) also reviewed the 2009 study and concluded that it "..presents no admissible scientific element likely to ascribe any haematological, hepatic or renal toxicity to the three re-analysed GMOs." The HCB also questioned the author's independence. Food Standards Australia New Zealand concluded that the results from the 2009 Séralini et al. study were due to chance alone.

Contamination Issues

In the 1990s genetically modified Flax tolerant to herbicide residues in soil was developed by the Crop Development Centre (CDC) at the University of Saskatchewan in Canada. Named Flax variety FP967, but commonly called *CDC Triffid*, research was controversially halted following protests from Canadian farmers who stood to lose up to 70% of their traditional export markets if it was introduced. GM Flax was deregistered, its sale was criminalized and in 2001 all modified seeds were destroyed. No modified crops had been planted and no seed had been sold but GM industry proponent Alan McHughen controversially passed out sample packets of seeds at presentations. In early September 2009, Flax imported into Germany was found to be contaminated with *CDC Triffid* causing the price of Canadian Flax to fall 32 percent. By mid November 35 countries reported contamination of imported Canadian Flax which has now been banned by the European Union. Canadian farmers are expected to be responsible for the cost of the cleanup and testing of future crops. In 2000, Aventis StarLink corn, which had been approved only as animal feed due to concerns about possible allergic reactions in humans, was found contaminating corn products in U.S. supermarkets. An episode involving Taco Bell taco shells was particularly well publicized which resulted in sales of StarLink seed being discontinued. The registration for the Starlink varieties was voluntarily withdrawn by Aventis in October 2000. Aid sent by the UN and the US to Central African nations was also found to be contaminated with StarLink corn and the aid was rejected. The US corn supply has been monitored for Starlink Bt proteins since 2001 and no positive samples have been found since 2004.

Chapter 6

Genetic Engineering in Animals

Genome organization refers to the sequential, not the structural organization of the genome. Besides the coding exons, the non-coding DNA in Eukaryotes may fall in the following classes:

- Introns. They are DNA sequences inserted between the exons and found in the ORF. They are spliced after the first level of transcription. Most introns are junk inserted within genes.
- Pseudogenes. 'Dead', non-functional copies of genes present elsewhere in the genome, but no longer of any use.
- Retropseudogenes. Like pseudogenes, but have been processed, i.e. lack introns. Produced by the action of reverse transcriptase (RT) on mRNA, and subsequent incorporation of the cDNA into the genome.
- Transposons. Jumping genes, which splice themselves in and out of the genome (in DNA form) randomly, by the action of transposase.
- Retrotransposons. Transcribed into an mRNA, which encodes an RT enzyme, which then copies the mRNA back to DNA and incorporates it into the genome.

Infact in humans only 1.5% of the entire genome length corresponds to coding DNA. This 1.5% codes for about 27,000 genes which in turn code for proteins that are responsible for all the cellular processes.

Genome Analysis?

Genome analysis entails the prediction of genes in uncharacterized genomic sequences. The 21st century has seen the announcement of the draft version of the human genome sequence. Model organisms have been sequenced in both the plant and animal kingdoms. However, the pace of genome annotation is not matching the pace of genome sequencing. Experimental genome annotation is slow and time consuming. The demand is to be able to develop computational tools for gene prediction. Computational Gene prediction is relatively simple for the prokaryotes where all the genes are converted into the corresponding mRNA and then into proteins. The process is more complex for eukaryotic cells where the coding DNA sequence is interrupted by random sequences called introns.

Some of the questions which biologists want to answer today are:

- Given a DNA sequence, what part of it codes for a protein and what part of it is junk DNA.
- Classify the junk DNA as intron, untranslated region, transposons, dead genes, regulatory elements etc.
- Divide a newly sequenced genome into the genes (coding) and the non-coding regions.

The importance of genome analysis can be understood by comparing the human and chimpanzee genomes. The chimp and human genomes vary by an average of just 2% i.e. just about 160 enzymes. A complete genome analysis of the two genomes would give a strong insight into the various mechanisms responsible for the differences. Arabidopsis and Humans have the same number of genes, though the Arabidopsis genome is around 250 times smaller than humans. How is that ? The human genome has a lot of junk DNA, specifically transposons and mobile genetic elements. This increases the size of the human genome, though the number of genes is only 27,000.

However, the number of protein products in humans is significantly higher. Many of the sequenced human genes have alternative splice products. In addition, several other processes (e.g. signal transduction) proceed via further protein modifications, such as Glycosylation. Therefore, the number of human protein products could far exceed the number of genes. Why do plants have such bulky genomes when they are not as complex as some of the higher eukaryotes ?

This is mostly due to two factors: the ability of plants to duplicate their genomes in order to reproduce (a process known as polyploidization) and the susceptibility of plants to mobile genetic elements.

- Polyploidization allows plants to more easily form hybrids when pollen and ova from different species come together. The result of such hybridization events are plants with genomes that are the sum of the two parent genome sizes (as opposed to half of one parent's genome and half of the other parent's genome as in normal sexual reproduction.
- Also, in case of plants, it is fairly common to observe insertion of transposable elements in intergenic regions. This also explains the difference in the sizes of plant genomes among themselves as well.

Open Reading Frame

The region of the nucleotide sequences from the start codon (ATG) to the stop codon is called the Open Reading frame. Gene finding in organism specially prokaryotes starts form searching for an open reading frames (ORF). An ORF is a sequence of DNA that starts with start codon "ATG" (not always) and ends with any of the three termination codons (TAA, TAG, TGA). Depending on the starting point, there are six possible ways (three on forward strand and three on complementary strand) of translating any nucleotide sequence into amino acid sequence according to the genetic code.These are called reading frames. While eukaryotic gene finding is altogether a different task as the eukaryotic genes are not continuous and interrupted by intervening noncoding sequences called 'introns'. Moreover organization of genetic information in eukaryotes and prokaryotes is different.

What is Coding Sequence

The Coding Sequence (CDS) is the actual region of DNA that is translated to form proteins. While the ORF may contain introns as well, the CDS refers to those nucleotides(concatenated exons) that can be divided into codons which are actually translated into amino acids by the ribosomal translation machinery. In Prokaryotes the ORF and the CDS are the same.

DNA Replication

Before a cell divides, its DNA is replicated (duplicated.) Because the two strands of a DNA molecule have complementary base pairs, the nucleotide sequence of each strand automatically supplies the information needed to produce its partner. If the two strands of a DNA molecule are separated, each can be used as a pattern or template to produce a complementary strand. Each template and its new complement together then form a new DNA double helix, identical to the original.

Before replication can occur, the length of the DNA double helix about to be copied must be unwound. In addition, the two strands must be separated, much like the two sides of a zipper, by breaking the weak hydrogen bonds that link the paired bases. Once the DNA strands have been unwound, they must be held apart to expose the bases so that new nucleotide partners can hydrogen-bond to them. The enzyme DNA polymerase then moves along the exposed DNA strand, joining newly arrived nucleotides into a new DNA strand that is complementary to the template. Each cell contains a family of more than thirty enzymes to insure the accurate replication of DNA.

Primers

Though DNA polymerase can elongate a polynucleotide strand by adding new nucleotides, it cannot start a strand from scratch because it can only bond new nucleotides to a free sugar (3') end of a nucleotide chain. DNA polymerase requires the assistance of a primer, a previously existing short strand of DNA (or RNA) that is complementary to the first part of the DNA segment being copied. This small strand of nucleotides anneals (binds) by complementary base pairing to the beginning of the area being copied. With the primer in place, DNA polymerase is then able to continue adding the rest of the pairs of the segment until a new double strand of DNA is completed. Primers are formed from free nucleotides in the cell by enzymes called DNA primases.

Replication Occurs Differently on Antiparallel Strands of DNA

That nucleotides can be added only to the sugar or 3' end of the growing complementary chain presents no problem for the side of the DNA chain opening at its phosphate or 5' end. The primer that binds to the first few exposed bases will end with a sugar (3') where the phosphate of a new nucleotide can be attached. From there on, DNA polymerase can continuously synthesize the growing complementary strand. This strand of DNA is called the leading strand. A different challenge faces DNA polymerase when the complementary side of the DNA molecule begins unzipping from its sugar (3') toward its phosphate (5') end. A primer of complementary molecules attaching to the opening end of this chain would have a phosphate not a sugar at its exposed end so that new nucleotides could not be joined.

To get around this problem, this strand is synthesized in small pieces backward from the overall direction of replication. This strand is called the lagging strand. The short segments of newly assembled DNA from which the lagging strand is built are called Okazaki fragments. As replication proceeds and nucleotides are added to the 3' end of the Okazaki fragments, they come to meet each other. The primer fragments are then booted out by enzymes and replaced by appropriate DNA nucleotides. The whole thing is then stitched together by another enzyme called DNA ligase.

DNA Damage, Repair, and Recombination

DNA repair refers to a collection of processes by which a cell identifies and corrects damage to the DNA molecules that encode its genome. In human cells, both normal metabolic activities and environmental factors such as UV light and radiation can cause DNA damage, resulting in as many as 1 million individual molecular lesions per cell per day.

Many of these lesions cause structural damage to the DNA molecule and can alter or eliminate the cell's ability to transcribe the gene that the affected DNA encodes. Other lesions induce potentially harmful mutations in the cell's genome, which affect the survival of its daughter cells after it undergoes mitosis.

As a consequence, the DNA repair process is constantly active as it responds to damage in the DNA structure. When normal repair processes fail, and when cellular apoptosis does not occur, irreparable DNA damage may occur, including double-strand breaks and DNA crosslinkages.

The rate of DNA repair is dependent on many factors, including the cell type, the age of the cell, and the extracellular environment. A cell that has accumulated a large amount of DNA damage, or one that no longer effectively repairs damage incurred to its DNA, can enter one of three possible states:

1. an irreversible state of dormancy, known as senescence
2. cell suicide, also known as apoptosis or programmed cell death
3. unregulated cell division, which can lead to the formation of a tumor that is cancerous.

The DNA repair ability of a cell is vital to the integrity of its genome and thus to its normal functioning and that of the organism. Many genes that were initially shown to influence life span have turned out to be involved in DNA damage repair and protection. Failure to correct molecular lesions in cells that form gametes can introduce mutations into the genomes of the offspring and thus influence the rate of evolution.

DNA Damage

DNA damage, due to environmental factors and normal metabolic processes inside the cell, occurs at a rate of 1,000 to 1,000,000 molecular lesions per cell per day. While this constitutes only 0.000165% of the human genome's approximately 6 billion bases (3 billion base pairs), unrepaired lesions in critical genes (such as tumor suppressor genes) can impede a cell's ability to carry out its function and appreciably increase the likelihood of tumour formation.

The vast majority of DNA damage affects the primary structure of the double helix; that is, the bases themselves are chemically modified. These modifications can in turn disrupt the molecules' regular helical structure by introducing non-native chemical bonds or bulky adducts that do not fit in the standard double helix.

Unlike proteins and RNA, DNA usually lacks tertiary structure and therefore damage or disturbance does not occur at that level. DNA is, however, supercoiled and wound around "packaging" proteins called histones (in eukaryotes), and both superstructures are vulnerable to the effects of DNA damage.

Sources of Damage

DNA damage can be subdivided into two main types:

1. endogenous damage such as attack by reactive oxygen species produced from normal metabolic byproducts (spontaneous mutation), especially the process of oxidative deamination
   - also includes replication errors.
2. exogenous damage caused by external agents such as
   - ultraviolet [UV 200-300nm] radiation from the sun
   - other radiation frequencies, including x-rays and gamma rays
   - hydrolysis or thermal disruption
   - certain plant toxins
   - human-made mutagenic chemicals, especially aromatic compounds that act as DNA intercalating agents
   - cancer chemotherapy and radiotherapy
   - viruses.

The replication of damaged DNA before cell division can lead to the incorporation of wrong bases opposite damaged ones. Daughter cells that inherit these wrong bases carry mutations from which the original DNA sequence is unrecoverable (except in the rare case of a back mutation, for example, through gene conversion).

Types of Damage

There are five main types of damage to DNA due to endogenous cellular processes:

1. *oxidation* of bases [e.g. 8-oxo-7,8-dihydroguanine (8-oxoG)] and generation of DNA strand interruptions from reactive oxygen species,
2. *alkylation* of bases (usually methylation), such as formation of 7-methylguanine, 1-methyladenine, 6-O-Methylguanine
3. *hydrolysis* of bases, such as deamination, depurination, and depyrimidination.
4. "bulky adduct formation" (i.e., benzo[a]pyrene diol epoxide-dG adduct)

5. *mismatch* of bases, due to errors in DNA replication, in which the wrong DNA base is stitched into place in a newly forming DNA strand, or a DNA base is skipped over or mistakenly inserted.

Damage caused by exogenous agents comes in many forms. Some examples are:

1. *UV-B light* causes crosslinking between adjacent cytosine and thymine bases creating *pyrimidine dimers*. This is called direct DNA damage.
2. *UV-A light* creates mostly free radicals. The damage caused by free radicals is called indirect DNA damage.
3. *Ionizing radiation* such as that created by radioactive decay or in *cosmic rays* causes breaks in DNA strands. Low-level ionizing radiation may induce irreparable DNA damage (leading to replicational and transcriptional errors needed for neoplasia or may trigger viral interactions) leading to pre-mature aging and cancer.
4. *Thermal disruption* at elevated temperature increases the rate of depurination (loss of purine bases from the DNA backbone) and single-strand breaks. For example, hydrolytic depurination is seen in the thermophilic bacteria, which grow in hot springs at 40-80 °C. The rate of depurination (300 purine residues per genome per generation) is too high in these species to be repaired by normal repair machinery, hence a possibility of an adaptive response cannot be ruled out.
5. *Industrial chemicals* such as vinyl chloride and hydrogen peroxide, and environmental chemicals such as polycyclic hydrocarbons found in smoke, soot and tar create a huge diversity of DNA adducts- ethenobases, oxidized bases, alkylated phosphotriesters and Crosslinking of DNA just to name a few.

UV damage, alkylation/methylation, X-ray damage and oxidative damage are examples of induced damage. Spontaneous damage can include the loss of a base, deamination, sugar ring puckering and tautomeric shift.

Nuclear versus Mitochondrial DNA Damage

In human cells, and eukaryotic cells in general, DNA is found in two cellular locations - inside the nucleus and inside the mitochondria. Nuclear DNA (nDNA) exists as chromatin during non-replicative stages of the cell cycle and is condensed into aggregate structures known as chromosomes during cell division. In either state the DNA is highly compacted and wound up around bead-like proteins called histones. Whenever a cell needs to express the genetic information encoded in its nDNA the required chromosomal region is unravelled, genes located therein are expressed, and then the region is condensed back to its resting conformation.

Mitochondrial DNA (mtDNA) is located inside mitochondria organelles, exists in multiple copies, and is also tightly associated with a number of proteins to form a complex known as the nucleoid. Inside mitochondria, reactive oxygen species (ROS), or free radicals, byproducts of the constant production of adenosine triphosphate (ATP) via oxidative phosphorylation, create a highly oxidative environment that is known to damage mtDNA. A critical enzyme in counteracting the toxicity of these species is superoxide dismutase, which is present in both the mitochondria and cytoplasm of eukaryotic cells.

Senescence and Apoptosis

Senescence, an irreversible state in which the cell no longer divides, is a protective response to the shortening of the chromosome ends. The telomeres are long regions of repetitive noncoding DNA that cap chromosomes and undergo partial degradation each time a cell undergoes division. In contrast, quiescence is a reversible state of cellular dormancy that is unrelated to genome damage. Senescence in cells may serve as a functional alternative to apoptosis in cases where the physical presence of a cell for spatial reasons is required by the organism, which serves as a "last resort" mechanism to prevent a cell with damaged DNA from replicating inappropriately in the absence of pro-growth cellular signaling. Unregulated cell division can lead to the formation of a tumour, which is potentially lethal to an organism. Therefore, the induction of senescence and apoptosis is considered to be part of a strategy of protection against cancer.

DNA Damage and Mutation

It is important to distinguish between DNA damage and mutation, the two major types of error in DNA. DNA damages and mutation are fundamentally different. Damages are physical abnormalities in the DNA, such as single- and double-strand breaks, 8-hydroxydeoxyguanosine residues, and polycyclic aromatic hydrocarbon adducts. DNA damages can be recognized by enzymes, and, thus, they can be correctly repaired if redundant information, such as the undamaged sequence in the complementary DNA strand or in a homologous chromosome, is available for copying. If a cell retains DNA damage, transcription of a gene can be prevented, and, thus, translation into a protein will also be blocked. Replication may also be blocked and/or the cell may die.

In contrast to DNA damage, a mutation is a change in the base sequence of the DNA. A mutation cannot be recognized by enzymes once the base change is present in both DNA strands, and, thus, a mutation cannot be repaired. At the cellular level, mutations can cause alterations in protein function and regulation. Mutations are replicated when the cell replicates. In a population of cells, mutant cells will increase or decrease in frequency according to the effects of the mutation on the ability of the cell to survive and reproduce. Although distinctly different from each other, DNA damages and mutations are related because DNA damages often cause errors of DNA synthesis during replication or repair; these errors are a major source of mutation. Given these properties of DNA damage and mutation, it can be seen that DNA damages are a special problem in non-dividing or slowly dividing cells, where unrepaired damages will tend to accumulate over time. On the other hand, in rapidly dividing cells, unrepaired DNA damages that do not kill the cell by blocking replication will tend to cause replication errors and thus mutation. The great majority of mutations that are not neutral in their effect are deleterious to a cell's survival.

Thus, in a population of cells comprising a tissue with replicating cells, mutant cells will tend to be lost. However, infrequent mutations that provide a survival advantage will tend to clonally expand at the expense of neighboring cells in the tissue. This advantage to the cell is disadvantageous to the whole organism, because such mutant cells can give rise to cancer. Thus, DNA damages in frequently dividing cells, because they give rise to mutations, are a prominent cause of cancer.

In contrast, DNA damages in infrequently dividing cells are likely a prominent cause of aging.

DNA Repair Mechanisms

Cells cannot function if DNA damage corrupts the integrity and accessibility of essential information in the genome (but cells remain superficially functional when so-called "non-essential" genes are missing or damaged). Depending on the type of damage inflicted on the DNA's double helical structure, a variety of repair strategies have evolved to restore lost information. If possible, cells use the unmodified complementary strand of the DNA or the sister chromatid as a template to recover the original information. Without access to a template, cells use an error-prone recovery mechanism known as translesion synthesis as a last resort. Damage to DNA alters the spatial configuration of the helix, and such alterations can be detected by the cell. Once damage is localized, specific DNA repair molecules bind at or near the site of damage, inducing other molecules to bind and form a complex that enables the actual repair to take place.

Direct Reversal

Cells are known to eliminate three types of damage to their DNA by chemically reversing it. These mechanisms do not require a template, since the types of damage they counteract can occur in only one of the four bases. Such direct reversal mechanisms are specific to the type of damage incurred and do not involve breakage of the phosphodiester backbone. The formation of pyrimidine dimers upon irradiation with UV light results in an abnormal covalent bond between adjacent pyrimidine bases.

The photoreactivation process directly reverses this damage by the action of the enzyme photolyase, whose activation is obligately dependent on energy absorbed from blue/UV light (300–500 nm wavelength) to promote catalysis. Another type of damage, methylation of guanine bases, is directly reversed by the protein methyl guanine methyl transferase (MGMT), the bacterial equivalent of which is called ogt.

This is an expensive process because each MGMT molecule can be used only once; that is, the reaction is stoichiometric rather than catalytic. A generalized response to methylating agents in bacteria is known as the adaptive response and confers a level of resistance to alkylating agents upon sustained exposure by upregulation of alkylation repair enzymes. The third type of DNA damage reversed by cells is certain methylation of the bases cytosine and adenine.

Single-strand Damage

When only one of the two strands of a double helix has a defect, the other strand can be used as a template to guide the correction of the damaged strand. In order to repair damage to one of the two paired molecules of DNA, there exist a number of excision repair mechanisms that remove the damaged nucleotide and replace it with an undamaged nucleotide complementary to that found in the undamaged DNA strand.

1. Base excision repair (BER), which repairs damage to a single base caused by oxidation, alkylation, hydrolysis, or deamination. The damaged base is removed by a DNA glycosylase. The "missing tooth" is then recognised by an enzyme called AP endonuclease, which cuts the Phosphodiester bond. The missing part is then resynthesized by a DNA polymerase, and a DNA ligase performs the final nick-sealing step.
2. Nucleotide excision repair (NER), which recognizes bulky, helix-distorting lesions such as pyrimidine dimers and 6,4 photoproducts. A specialized form of NER known as transcription-coupled repair deploys NER enzymes to genes that are being actively transcribed.
3. Mismatch repair (MMR), which corrects errors of DNA replication and recombination that result in mispaired (but undamaged) nucleotides.

Double-strand Breaks

Double-strand breaks, in which both strands in the double helix are severed, are particularly hazardous to the cell because they can lead to genome rearrangements. Three mechanisms exist to repair double-strand breaks (DSBs): non-homologous end joining (NHEJ), microhomology-mediated end joining (MMEJ), and homologous recombination.

PVN Acharya noted that double-strand breaks and a "cross-linkage joining both strands at the same point is irreparable because neither strand can then serve as a template for repair. The cell will die in the next mitosis or in some rare instances, mutate." In NHEJ, DNA Ligase IV, a specialized DNA ligase that forms a complex with the cofactor XRCC4, directly joins the two ends. To guide accurate repair, NHEJ relies on short homologous sequences called microhomologies present on the single-stranded tails of the DNA ends to be joined. If these overhangs are compatible, repair is usually accurate.

NHEJ can also introduce mutations during repair. Loss of damaged nucleotides at the break site can lead to deletions, and joining of nonmatching termini forms translocations. NHEJ is especially important before the cell has replicated its DNA, since there is no template available for repair by homologous recombination. There are "backup" NHEJ pathways in higher eukaryotes. Besides its role as a genome caretaker, NHEJ is required for joining hairpin-capped double-strand breaks induced during V(D)J recombination, the process that generates diversity in B-cell and T-cell receptors in the vertebrate immune system.

Homologous recombination requires the presence of an identical or nearly identical sequence to be used as a template for repair of the break. The enzymatic machinery responsible for this repair process is nearly identical to the machinery responsible for chromosomal crossover during meiosis. This pathway allows a damaged chromosome to be repaired using a sister chromatid (available in G2 after DNA replication) or a homologous chromosome as a template. DSBs caused by the replication machinery attempting to synthesize across a single-strand break or unrepaired lesion cause collapse of the replication fork and are typically repaired by recombination. Topoisomerases introduce both single- and double-strand breaks in the course of changing the DNA's state of supercoiling, which is especially common in regions near an open replication fork. Such breaks are not considered DNA damage because they are a natural intermediate in the topoisomerase biochemical mechanism and are immediately repaired by the enzymes that created them.

A team of French researchers bombarded *Deinococcus radiodurans* to study the mechanism of double-strand break DNA repair in that organism. At least two copies of the genome, with random DNA breaks, can form DNA fragments through annealing. Partially overlapping fragments are then used for synthesis of homologous regions through a moving D-loop that can continue extension until they find complementary partner strands. In the final step there is crossover by means of RecA-dependent homologous recombination.

Translesion Synthesis

Translesion synthesis is a DNA damage tolerance process that allows the DNA replication machinery to replicate past DNA lesions such as thymine dimers or AP sites. It involves switching out regular DNA polymerases for specialized translesion polymerases (i.e. DNA polymerase IV or V, from the Y Polymerase family), often with larger active sites that can facilitate the insertion of bases opposite damaged nucleotides. The polymerase switching is thought to be mediated by, among other factors, the post-translational modification of the replication processivity factor PCNA.

Translesion synthesis polymerases often have low fidelity (high propensity to insert wrong bases) on undamaged templates relative to regular polymerases. However, many are extremely efficient at inserting correct bases opposite specific types of damage. For example, Pol ç mediates error-free bypass of lesions induced by UV irradiation, whereas Pol æ introduces mutations at these sites.

From a cellular perspective, risking the introduction of point mutations during translesion synthesis may be preferable to resorting to more drastic mechanisms of DNA repair, which may cause gross chromosomal aberrations or cell death. In short, the process involves specialized polymerases either bypassing or repairing lesions at locations of stalled DNA replication. A bypass platform is provided to these polymerases by Proliferating cell nuclear antigen (PCNA). Under normal circumstances, PCNA bound to polymerases replicates the DNA. At a site of lesion, PCNA is ubiquitinated, or modified, by the RAD6/RAD18 proteins to provide a platform for the specialized polymerases to bypass the lesion and resume DNA replication.

Global Response to DNA Damage

Cells exposed to ionizing radiation, ultraviolet light or chemicals are prone to acquire multiple sites of bulky DNA lesions and double-strand breaks. Moreover, DNA damaging agents can damage other biomolecules such as proteins, carbohydrates, lipids, and RNA. The accumulation of damage, to be specific, double-strand breaks or adducts stalling the replication forks, are among known stimulation signals for a global response to DNA damage. The global response to damage is an act directed toward the cells' own preservation and triggers multiple pathways of macromolecular repair, lesion bypass, tolerance, or apoptosis. The common features of global response are induction of multiple genes, cell cycle arrest, and inhibition of cell division.

DNA Damage Checkpoints

After DNA damage, cell cycle checkpoints are activated. Checkpoint activation pauses the cell cycle and gives the cell time to repair the damage before continuing to divide. DNA damage checkpoints occur at the G1/S and G2/M boundaries. An intra-S checkpoint also exists. Checkpoint activation is controlled by two master kinases, ATM and ATR. ATM responds to DNA double-strand breaks and disruptions in chromatin structure, whereas ATR primarily responds to stalled replication forks. These kinases phosphorylate downstream targets in a signal transduction cascade, eventually leading to cell cycle arrest. A class of checkpoint mediator proteins including BRCA1, MDC1, and 53BP1 has also been identified. These proteins seem to be required for transmitting the checkpoint activation signal to downstream proteins.

P53 is an important downstream target of ATM and ATR, as it is required for inducing apoptosis following DNA damage. At the G1/S checkpoint, p53 functions by deactivating the CDK2/cyclin E complex. Similarly, p21 mediates the G2/M checkpoint by deactivating the CDK1/cyclin B complex.

The Prokaryotic SOS Response

The SOS response is the term used to describe changes in gene expression in *Escherichia coli* and other bacteria in response to extensive DNA damage. The prokaryotic SOS system is regulated by two key proteins: LexA and RecA. The LexA homodimer is a transcriptional repressor that binds to operator sequences commonly referred to as SOS boxes.

In *Escherichia coli* it is known that LexA regulates transcription of approximately 48 genes including the lexA and recA genes. The SOS response is known to be widespread in the Bacteria domain, but it is mostly absent in some bacterial phyla, like the Spirochetes. The most common cellular signals activating the SOS response are regions of single-stranded DNA (ssDNA), arising from stalled replication forks or double-strand breaks, which are processed by DNA helicase to separate the two DNA strands.

In the initiation step, RecA protein binds to ssDNA in an ATP hydrolysis driven reaction creating RecA–ssDNA filaments. RecA–ssDNA filaments activate LexA autoprotease activity, which ultimately leads to cleavage of LexA dimer and subsequent LexA degradation. The loss of LexA repressor induces transcription of the SOS genes and allows for further signal induction, inhibition of cell division and an increase in levels of proteins responsible for damage processing.

In *Escherichia coli*, SOS boxes are 20-nucleotide long sequences near promoters with palindromic structure and a high degree of sequence conservation. In other classes and phyla, the sequence of SOS boxes varies considerably, with different length and composition, but it is always highly conserved and one of the strongest short signals in the genome. The high information content of SOS boxes permits differential binding of LexA to different promoters and allows for timing of the SOS response. The lesion repair genes are induced at the beginning of SOS response. The error-prone translesion polymerases, for example, UmuCD'2 (also called DNA polymerase V), are induced later on as a last resort. Once the DNA damage is repaired or bypassed using polymerases or through recombination, the amount of single-stranded DNA in cells is decreased, lowering the amounts of RecA filaments decreases cleavage activity of LexA homodimer, which then binds to the SOS boxes near promoters and restores normal gene expression.

Eukaryotic Transcriptional Responses to DNA Damage

Eukaryotic cells exposed to DNA damaging agents also activate important defensive pathways by inducing multiple proteins involved in DNA repair, cell cycle checkpoint control, protein trafficking and degradation. Such genome wide transcriptional response is very complex and tightly regulated, thus allowing coordinated global response to damage. Exposure of yeast *Saccharomyces cerevisiae* to DNA damaging agents results in overlapping but distinct transcriptional profiles. Similarities to environmental shock response indicates that a general global stress response pathway exist at the level of transcriptional activation. In contrast, different human cell types respond to damage differently indicating an absence of a common global response. The probable explanation for this difference between yeast and human cells may be in the heterogeneity of mammalian cells. In an animal different types of cells are distributed among different organs that have evolved different sensitivities to DNA damage.

In general global response to DNA damage involves expression of multiple genes responsible for postreplication repair, homologous recombination, nucleotide excision repair, DNA damage checkpoint, global transcriptional activation, genes controlling mRNA decay, and many others. A large amount of damage to a cell leaves it with an important decision: undergo apoptosis and die, or survive at the cost of living with a modified genome. An increase in tolerance to damage can lead to an increased rate of survival that will allow a greater accumulation of mutations. Yeast Rev1 and human polymerase ç are members of [Y family translesion DNA polymerases present during global response to DNA damage and are responsible for enhanced mutagenesis during a global response to DNA damage in eukaryotes.

DNA Repair and Aging

DNA repair rate is an important determinant of cell pathology. Experimental animals with genetic deficiencies in DNA repair often show decreased life span and increased cancer incidence. For example, mice deficient in the dominant NHEJ pathway and in telomere maintenance mechanisms get lymphoma and infections more often, and, as a consequence, have shorter lifespans than wild-type mice. In similar manner, mice deficient in a key repair and transcription protein that unwinds DNA helices have premature onset of aging-related diseases and consequent shortening of lifespan. However, not every DNA repair deficiency creates exactly the predicted effects;

mice deficient in the NER pathway exhibited shortened life span without correspondingly higher rates of mutation.

If the rate of DNA damage exceeds the capacity of the cell to repair it, the accumulation of errors can overwhelm the cell and result in early senescence, apoptosis, or cancer. Inherited diseases associated with faulty DNA repair functioning result in premature aging, increased sensitivity to carcinogens, and correspondingly increased cancer risk. On the other hand, organisms with enhanced DNA repair systems, such as *Deinococcus radiodurans*, the most radiation-resistant known organism, exhibit remarkable resistance to the double-strand break-inducing effects of radioactivity, likely due to enhanced efficiency of DNA repair and especially NHEJ.

Longevity and Caloric Restriction

A number of individual genes have been identified as influencing variations in life span within a population of organisms. The effects of these genes is strongly dependent on the environment, in particular, on the organism's diet. Caloric restriction reproducibly results in extended lifespan in a variety of organisms, likely via nutrient sensing pathways and decreased metabolic rate. The molecular mechanisms by which such restriction results in lengthened lifespan are as yet unclear; however, the behavior of many genes known to be involved in DNA repair is altered under conditions of caloric restriction. For example, increasing the gene dosage of the gene SIR-2, which regulates DNA packaging in the nematode worm *Caenorhabditis elegans*, can significantly extend lifespan. The mammalian homolog of SIR-2 is known to induce downstream DNA repair factors involved in NHEJ, an activity that is especially promoted under conditions of caloric restriction. Caloric restriction has been closely linked to the rate of base excision repair in the nuclear DNA of rodents, although similar effects have not been observed in mitochondrial DNA.

It is interesting to note that the *C. elegans* gene AGE-1, an upstream effector of DNA repair pathways, confers dramatically extended life span under free-feeding conditions but leads to a decrease in reproductive fitness under conditions of caloric restriction. This observation supports the pleiotropy theory of the biological origins of aging, which suggests that genes conferring a large survival advantage early in life will be selected for even if they carry a corresponding disadvantage late in life.

Medicine and DNA Repair Modulation

Defects in the NER mechanism are responsible for several genetic disorders, including:

- *Xeroderma pigmentosum:* hypersensitivity to sunlight/UV, resulting in increased skin cancer incidence and premature aging
- *Cockayne syndrome:* hypersensitivity to UV and chemical agents
- Trichothiodystrophy: sensitive skin, brittle hair and nails.

Mental retardation often accompanies the latter two disorders, suggesting increased vulnerability of developmental neurons.

Other DNA repair disorders include:

- Werner's syndrome: premature aging and retarded growth
- Bloom's syndrome: sunlight hypersensitivity, high incidence of malignancies (especially leukemias).
- Ataxia telangiectasia: sensitivity to ionizing radiation and some chemical agents.

All of the above diseases are often called "segmental progerias" ("accelerated aging diseases") because their victims appear elderly and suffer from aging-related diseases at an abnormally young age, while not manifesting all the symptoms of old age. Other diseases associated with reduced DNA repair function include Fanconi's anemia, hereditary breast cancer and hereditary colon cancer.

DNA Repair and Cancer

Inherited mutations that affect DNA repair genes are strongly associated with high cancer risks in humans. Hereditary nonpolyposis colorectal cancer (HNPCC) is strongly associated with specific mutations in the DNA mismatch repair pathway. BRCA1 and BRCA2, two famous mutations conferring a hugely increased risk of breast cancer on carriers, are both associated with a large number of DNA repair pathways, especially NHEJ and homologous recombination. Cancer therapy procedures such as chemotherapy and radiotherapy work by overwhelming the capacity of the cell to repair DNA damage, resulting in cell death. Cells that are most rapidly dividing - most typically cancer cells - are preferentially affected. The side-effect is that other non-cancerous but rapidly dividing cells such as stem cells in the bone marrow are also affected. Modern cancer treatments attempt to localize the DNA damage to cells and tissues only associated with cancer, either by physical means (concentrating the therapeutic agent in the region of the tumour) or by biochemical means (exploiting a feature unique to cancer cells in the body).

DNA Repair and Evolution

The basic processes of DNA repair are highly conserved among both prokaryotes and eukaryotes and even among bacteriophage (viruses that infect bacteria); however, more complex organisms with more complex genomes have correspondingly more complex repair mechanisms. The ability of a large number of protein structural motifs to catalyze relevant chemical reactions has played a significant role in the elaboration of repair mechanisms during evolution. For an extremely detailed review of hypotheses relating to the evolution of DNA repair. The fossil record indicates that single-cell life began to proliferate on the planet at some point during the Precambrian period, although exactly when recognizably modern life first emerged is unclear. Nucleic acids became the sole and universal means of encoding genetic information, requiring DNA repair mechanisms that in their basic form have been inherited by all extant life forms from their common ancestor.

The emergence of Earth's oxygen-rich atmosphere (known as the "oxygen catastrophe") due to photosynthetic organisms, as well as the presence of potentially damaging free radicals in the cell due to oxidative phosphorylation, necessitated the evolution of DNA repair mechanisms that act specifically to counter the types of damage induced by oxidative stress.

Rate of Evolutionary Change

On some occasions, DNA damage is not repaired, or is repaired by an error-prone mechanism that results in a change from the original sequence. When this occurs, mutations may propagate into the genomes of the cell's progeny. Should such an event occur in a germ line cell that will eventually produce a gamete, the mutation has the potential to be passed on to the organism's offspring. The rate of evolution in a particular species (or, in a particular gene) is a function of the rate of mutation. As a consequence, the rate and accuracy of DNA repair mechanisms have an influence over the process of evolutionary change.

Senescence

Senescence or biological aging is the change in the biology of an organism as it ages after its maturity. Such changes range from those affecting its cells and their function to that of the whole organism. There are a number of theories as to why senescence occurs, including ones that claim it is programmed by gene expression changes and that it is the accumulative damage of biological processes. The word *senescence* is derived from the Latin word *senex*, meaning *old man*, *old age*, or *advanced in age*.

Cellular Senescence

Cellular senescence is the phenomenon by which normal diploid cells lose the ability to divide, normally after about 50 cell divisions in vitro. Some cells become senescent after fewer replications cycles as a result of DNA double strand breaks, toxins, etc. This phenomenon is also known as "replicative senescence", the "Hayflick phenomenon", or the Hayflick limit in honour of Dr. Leonard Hayflick who was the first to publish this information in 1965.

In response to DNA damage (including shortened telomeres), cells either age or self-destruct (apoptosis, programmed cell death) if the damage cannot be repaired. In this 'cellular suicide', the death of one cell, or more, may benefit the organism as a whole. For example, in plants the death of the water-conducting xylem cells (tracheids and vessel elements) allows the cells to function more efficiently and so deliver water to the upper parts of a plant.

Aging of the Whole Organism

Organismal senescence is the aging of whole organisms. In general, aging is characterized by the declining ability to respond to stress, increased homeostatic imbalance, and increased risk of aging-associated diseases. Death is the ultimate consequence of aging, though "old age" is not a scientifically recognized cause of death because there is always a specific proximal cause, such as cancer, heart disease, or liver failure. Aging of whole organisms is therefore a complex process that can be defined as "a progressive deterioration of physiological function, an intrinsic age-related process of loss of viability and increase in vulnerability".

Differences in maximum life span among species correspond to different "rates of aging". For example, inherited differences in the rate of aging make a mouse elderly at 3 years and a human elderly at 80 years. These genetic differences affect a variety of physiological processes, including the efficiency of DNA repair, antioxidant enzymes, and rates of free radical production. Senescence of the organism gives rise to the Gompertz–Makeham law of mortality, which says that mortality rate rises rapidly with age. Some animals, such as some reptiles and fish, age slowly (negligible senescence) and exhibit very long lifespans. Some even exhibit "negative senescence", in which mortality falls with age, in disagreement with the Gompertz–Makeham "law". Whether replicative senescence (Hayflick limit) plays a causative role in organismal aging is at present an active area of investigation.

Theories of Aging

The process of senescence is complex, and may derive from a variety of different mechanisms and exist for a variety of different reasons. However, senescence is not universal, and scientific evidence suggests that cellular senescence evolved in certain species because it prevents the onset of cancer. In a few simple species, such as Hydra, senescence is negligible and cannot be detected. All such species have no "post-mitotic" cells; they reduce the effect of damaging free radicals by cell division and dilution. Such species are not immortal, however, as they will eventually fall prey to trauma or disease. Moreover, average lifespans can vary greatly within and between species. This suggests that both genetic and environmental factors contribute to aging.

In general, theories that explain senescence have been divided between the programmed and stochastic theories of aging. Programmed theories imply that aging is regulated by biological clocks operating throughout the lifespan. This regulation would depend on changes in gene expression that affect the systems responsible for maintenance, repair, and defense responses. Stochastic theories blame environmental impacts on living organisms that induce cumulative damage at various levels as the cause of aging, examples of which ranging from damage to DNA, damage to tissues and cells by oxygen radicals (widely known as free radicals countered by the even more well-known antioxidants), and cross-linking.

However, aging is seen as a progressive failure of homeodynamics (homeostasis) involving genes for the maintenance and repair, stochastic events leading to molecular damage and molecular heterogeneity, and chance events determining the probability of death. Since complex and interacting systems of maintenance and repair comprise the homeodynamic (old term: homeostasis) space of a biological system, aging is considered to be a progressive shrinkage of homeodynamic space mainly due to increased molecular heterogeneity.

Evolutionary Theories

A gene can be expressed at various life-stages. Therefore, natural selection can support lethal and harmful alleles, if their expression occurs after reproduction. Senescence may be the product of such selection. In addition, aging is believed to have evolved because of the increasingly smaller probability of an organism still being alive at older age, due to predation and accidents, both of which may be random and age-invariant. It is thought that strategies that result in a higher reproductive rate at a young age, but shorter overall lifespan, result in a higher lifetime reproductive success and are therefore favoured by natural selection. In essence, aging is, therefore, the result of investing resources in reproduction, rather than maintenance of the body (the "Disposable Soma" theory), in light of the fact that accidents, predation, and disease will eventually kill the organism no matter how much energy is devoted to repair of the body. Various other theories of aging exist, and are not necessarily mutually exclusive.

The geneticist J. B. S. Haldane wondered why the dominant mutation that causes Huntington's disease remained in the population, and why natural selection had not eliminated it. The onset of this neurological disease is (on average) at age 45 and is invariably fatal within 10–20 years. Haldane assumed that, in human prehistory, few survived until age 45.

Since few were alive at older ages and their contribution to the next generation was therefore small relative to the large cohorts of younger age groups, the force of selection against such late-acting deleterious mutations was correspondingly small. However, if a mutation affected younger individuals, selection against it would be strong. Therefore, late-acting deleterious mutations could accumulate in populations over evolutionary time through genetic drift. This principle has been demonstrated experimentally. And it is these later-acting deleterious mutations that are believed to cause—even allow—age-related mortality. Peter Medawar formalised this observation in his mutation accumulation theory of aging. "The force of natural selection weakens with increasing age—even in a theoretically immortal population, provided only that it is exposed to real hazards of mortality. If a genetic disaster... happens late enough in individual life, its consequences may be completely unimportant".

The 'real hazards of mortality' are, in typical circumstances, predation, disease, and accidents. So, even an immortal population, whose fertility does not decline with time, will have fewer individuals alive in older age groups. This is called 'extrinsic mortality'. Young cohorts, not depleted in numbers yet by extrinsic mortality, contribute far more to the next generation than the few remaining older cohorts, so the force of selection against late-acting deleterious mutations, which affect only these few older individuals, is very weak. The mutations may not be selected against, therefore, and may spread over evolutionary time into the population.

The major testable prediction made by this model is that species that have high extrinsic mortality in nature will age more quickly and have shorter intrinsic lifespans. This is borne out among mammals, the best-studied in terms of life history. There is a correlation among mammals between body size and lifespan, such that larger species live longer than smaller species in controlled/optimum conditions, but there are notable exceptions. For instance, many bats and rodents are of similar size, yet bats live much longer. For instance, the little brown bat, half the size of a mouse, can live 30 years in the wild. A mouse will live 2–3 years even with optimum conditions. The explanation is that bats have fewer predators, so therefore low extrinsic mortality. Thus, more individuals survive to later ages, so the force of selection against late-acting deleterious mutations is stronger. Fewer late-acting deleterious mutations = slower aging = longer lifespan. Birds are also warm-blooded and are of similar size to many small mammals, yet live often 5–10 times as long. They have fewer predation pressures compared with ground-dwelling mammals. And seabirds, which, in general, have the fewest predators of all birds, live longest.

Also, when examining the body-size vs. lifespan relationship, one will observe that predator mammals tend to have longer lifespans than prey animals in a controlled environment such as a zoo or nature reserve. The explanation for the long lifespans of primates (such as humans, monkeys, and apes) relative to body size is that their intelligence and often sociality helps them avoid becoming prey. Being a predator, being smart, and working together all reduce extrinsic mortality.

Another evolutionary theory of aging was proposed by George C. Williams and involves antagonistic pleiotropy. A single gene may affect multiple traits. Some traits that increase fitness early in life may also have negative effects later in life. But, because many more individuals are alive at young ages than at old ages, even small positive effects early can be strongly selected for, and large negative effects later may be very weakly selected against. Williams suggested the following example: Perhaps a gene codes for calcium deposition in bones, which promotes juvenile survival and will therefore be favored by natural selection; however, this same gene promotes calcium deposition in the arteries, causing negative effects in old age. Therefore, negative effects in old age may reflect the result of natural selection for pleiotropic genes that are beneficial early in life. In this case, fitness is relatively high when Fisher's reproductive value is high and relatively low when Fisher's reproductive value is low.

A number of genetic components of aging have been identified using model organisms, ranging from the simple budding yeast *Saccharomyces cerevisiae* to worms such as *Caenorhabditis elegans* and fruit flies (*Drosophila melanogaster*). Study of these organisms has revealed the presence of at least two conserved aging pathways. One of these pathways involves the gene *Sir2*, a NAD+-dependent histone deacetylase. In yeast, Sir2 is required for genomic silencing at three loci: The yeast mating loci, the telomeres and the ribosomal DNA (rDNA). In some species of yeast, replicative aging may be partially caused by homologous recombination between rDNA repeats; excision of rDNA repeats results in the formation of extrachromosomal rDNA circles (ERCs).

These ERCs replicate and preferentially segregate to the mother cell during cell division, and are believed to result in cellular senescence by titrating away (competing for) essential nuclear factors. ERCs have not been observed in other species (nor even all strains of the same yeast species) of yeast (which also display replicative senescence), and ERCs are not believed to contribute to aging in higher organisms such as humans (they have not been shown to accumulate in mammals in a similar manner to yeast). Extrachromosomal circular DNA (eccDNA) has been found in worms, flies, and humans. The origin and role of eccDNA in aging, if any, is unknown.

Despite the lack of a connection between circular DNA and aging in higher organisms, extra copies of Sir2 are capable of extending the lifespan of both worms and flies (though, in flies, this finding has not been replicated by other investigators, and the activator of Sir2 resveratrol does not reproducibly increase lifespan in either species). Whether the Sir2 homologues in higher organisms have any role in lifespan is unclear, but the human SIRT1 protein has been demonstrated to deacetylate p53, Ku70, and the forkhead family of transcription factors. SIRT1 can also regulate acetylates such as CBP/p300, and has been shown to deacetylate specific histone residues. RAS1 and RAS2 also affect aging in yeast and have a human homologue. RAS2 overexpression has been shown to extend lifespan in yeast. Other genes regulate aging in yeast by increasing the resistance to oxidative stress. Superoxide dismutase, a protein that protects against the effects of mitochondrial free radicals, can extend yeast lifespan in stationary phase when overexpressed.

In higher organisms, aging is likely to be regulated in part through the insulin/IGF-1 pathway. Mutations that affect insulin-like signaling in worms, flies, and the growth hormone/ IGF1 axis in mice are associated with extended lifespan. In yeast, Sir2 activity is regulated by the nicotinamidase PNC1. PNC1 is transcriptionally upregulated under stressful conditions such as caloric restriction, heat shock, and osmotic shock. By converting nicotinamide to niacin, nicotinamide is removed, inhibiting the activity of Sir2. A nicotinamidase found in humans, known as PBEF, may serve a similar function, and a secreted form of PBEF known as visfatin may help to regulate serum insulin levels. It is not known, however, whether these mechanisms also exist in humans, since there are obvious differences in biology between humans and model organisms.

Sir2 activity has been shown to increase under calorie restriction. Due to the lack of available glucose in the cells, more NAD+ is available and can activate Sir2. Resveratrol, a stilbenoid found in the skin of red grapes, was reported to extend the lifespan of yeast, worms, and flies (the lifespan extension in flies and worms have proved irreproducible by independent investigators). It has been shown to activate Sir2 and therefore mimics the effects of calorie restriction, if one accepts that caloric restriction is indeed dependent on Sir2.

Gene expression is imperfectly controlled, and it is possible that random fluctuations in the expression levels of many genes contribute to the aging process as suggested by a study of such genes in yeast. Individual cells, which are genetically identical, none-the-less can have substantially different responses to outside stimuli, and markedly different lifespans, indicating the epigenetic factors play an important role in gene expression and aging as well as genetic factors. According to the GenAge database of aging-related genes there are over 700 genes associated with aging in model organisms: 555 in the soil roundworm (*Caenorhabditis elegans*), 87 in the bakers' yeast (*Saccharomyces cerevisiae*), 75 in the fruit fly (*Drosophila melanogaster*) and 68 in the mouse (*Mus musculus*).

As noted above, senescence is not universal, and senescence is not observed in single-celled organisms that reproduce through the process of cellular mitosis. Moreover, cellular senescence is not observed in several organisms, including perennial plants, sponges, corals, and lobsters. In those species where cellular senescence is observed, cells eventually become post-mitotic when they can no longer replicate themselves through the process of cellular mitosis; i.e., cells experience *replicative senescence.*

How and why some cells become post-mitotic in some species has been the subject of much research and speculation, but (as noted above) it is widely believed that cellular senescence evolved as a way to prevent the onset and spread of cancer. Somatic cells that have divided many times will have accumulated DNA mutations and would therefore be in danger of becoming cancerous if cell division continued. Lately, the role of telomeres in cellular senescence has aroused general interest, especially with a view to the possible genetically adverse effects of cloning. The successive shortening of the chromosomal telomeres with each cell cycle is also believed to limit the number of divisions of the cell, thus contributing to aging.

There have, on the other hand, also been reports that cloning could alter the shortening of telomeres. Some cells do not age and are, therefore, described as being "biologically immortal". It is theorized by some that when it is discovered exactly what allows these cells, whether it be the result of telomere lengthening or not, to divide without limit that it will be possible to genetically alter other cells to have the same capability. It is further theorized that it will eventually be possible to genetically engineer all cells in the human body to have this capability by employing gene therapy and, therefore, stop or reverse aging, effectively making the entire organism potentially immortal. Cancer cells are usually immortal. In about 85% of tumors, this evasion of cellular senescence is the result of up-activation of their telomerase genes. This simple observation suggests that reactivation of telomerase in healthy individuals could greatly increase their cancer risk. Whether cell senescence plays any role in organismal aging is at present unknown, and is an active area of investigation. Mouse mutants lacking telomerase do not immediately show accelerated aging.

Chemical Damage

One of the earliest aging theories was the *Rate of Living Hypothesis* described by Raymond Pearl in 1928(based on earlier work by Max Rubner), which states that fast basal metabolic rate corresponds to short maximum life span.

While there may be some validity to the idea that for various types of specific damage detailed below that are by-products of metabolism, all other things being equal, a fast metabolism may reduce lifespan, in general this theory does not adequately explain the differences in lifespan either within, or between, species. Calorically-restricted animals process as much, or more, calories per gram of body mass, as their *ad libitum* fed counterparts, yet exhibit substantially longer lifespans.. Similarly, metabolic rate is a poor predictor of lifespan for birds, bats and other species that, it is presumed, have reduced mortality from predation, and therefore have evolved long lifespans even in the presence of very high metabolic rates. More recently, it was shown that, when modern statistical methods for correcting for the effects of body size and phylogeny are employed, metabolic rate does not correlate with longevity in mammals or birds.

With respect to specific types of chemical damage caused by metabolism, it is suggested that damage to long-lived biopolymers, such as structural proteins or DNA, caused by ubiquitous chemical agents in the body such as oxygen and sugars, are in part responsible for aging. The damage can include breakage of biopolymer chains, cross-linking of biopolymers, or chemical attachment of unnatural substituents (haptens) to biopolymers.

Under normal aerobic conditions, approximately 4% of the oxygen metabolized by mitochondria is converted to superoxide ion, which can subsequently be converted to hydrogen peroxide, hydroxyl radical and eventually other reactive species including other peroxides and singlet oxygen, which can, in turn, generate free radicals capable of damaging structural proteins and DNA. Certain metal ions found in the body, such as copper and iron, may participate in the process. (In Wilson's disease, a hereditary defect that causes the body to retain copper, some of the symptoms resemble accelerated senescence.) These processes are termed *oxidative damage* and are linked to the benefits of nutritionally derived polyphenol antioxidants. Sugars such as glucose and fructose can react with certain amino acids such as lysine and arginine and certain DNA bases such as guanine to produce sugar adducts, in a process called *glycation*.

These adducts can further rearrange to form reactive species, which can then cross-link the structural proteins or DNA to similar biopolymers or other biomolecules such as non-structural proteins. People with diabetes, who have elevated blood sugar, develop senescence-associated disorders much earlier than the general population, but can delay such disorders by rigorous control of their blood sugar levels. There is evidence that sugar damage is linked to oxidant damage in a process termed *glycoxidation*.

Free radicals can damage proteins, lipids or DNA. Glycation mainly damages proteins. Damaged proteins and lipids accumulate in lysosomes as lipofuscin. Chemical damage to structural proteins can lead to loss of function; for example, damage to collagen of blood vessel walls can lead to vessel-wall stiffness and, thus, hypertension, and vessel wall thickening and reactive tissue formation (atherosclerosis); similar processes in the kidney can lead to renal failure. Damage to enzymes reduces cellular functionality. Lipid peroxidation of the inner mitochondrial membrane reduces the electric potential and the ability to generate energy. It is probably no accident that nearly all of the so-called "accelerated aging diseases" are due to defective DNA repair enzymes.

It is believed that the impact of alcohol on aging can be partly explained by alcohol's activation of the HPA axis, which stimulates glucocorticoid secretion, long-term exposure to which produces symptoms of aging.

Reliability Theory

Reliability theory suggests that biological systems start their adult life with a high load of initial damage. Reliability theory is a general theory about systems failure. It allows researchers to predict the age-related failure kinetics for a system of given architecture (reliability structure) and given reliability of its components. Reliability theory predicts that even those systems that composed entirely of non-aging elements (with a constant failure rate) will nevertheless deteriorate (fail more often) with age, if these systems are redundant in irreplaceable elements. Aging, therefore, is a direct consequence of systems.

Reliability theory also predicts the late-life mortality deceleration with subsequent leveling-off, as well as the late-life mortality plateaus, as an inevitable consequence of redundancy exhaustion at extreme old ages. The theory explains why mortality rates increase exponentially with age (the Gompertz law) in many species, by taking into account the initial flaws (defects) in newly formed systems. It also explains why organisms "prefer" to die according to the Gompertz law, while technical devices usually fail according to the Weibull (power) law. Reliability theory allows to specify conditions when organisms die according to the Weibull distribution: Organisms should be relatively free of initial flaws and defects.

The theory makes it possible to find a general failure law applicable to all adult and extreme old ages, where the Gompertz and the Weibull laws are just special cases of this more general failure law. The theory explains why relative differences in mortality rates of compared populations (within a given species) vanish with age (compensation law of mortality), and mortality convergence is observed due to the exhaustion of initial differences in redundancy levels.

Miscellaneous

Recently, a kind of early senescence has been alleged to be a possible unintended outcome of early cloning experiments. The issue was raised in the case of Dolly the sheep, following her death from a contagious lung disease. The claim that Dolly's early death involved premature senescence has been vigorously contested, and Dolly's creator, Dr. Ian Wilmut has expressed the view that her illness and death were probably unrelated to the fact that she was a clone.

A set of rare hereditary (genetic) disorders, each called progeria, has been known for some time. Sufferers exhibit symptoms resembling accelerated aging, including wrinkled skin. The cause of Hutchinson–Gilford progeria syndrome was reported in the journal *Nature* in May 2003. This report suggests that DNA damage, not oxidative stress, is the cause of this form of accelerated aging.

Transcription

Transcription is the process of creating a complementary RNA copy of a sequence of DNA. Both RNA and DNA are nucleic acids, which use base pairs of nucleotides as a complementary language that can be converted back and forth from DNA to RNA by the action of the correct enzymes. During transcription, a DNA sequence is read by RNA polymerase, which produces a complementary, antiparallel RNA strand. As opposed to DNA replication, transcription results in an RNA complement that includes uracil (U) in all instances where thymine (T) would have occurred in a DNA complement.

Transcription can be explained easily in 4 or 5 simple steps, each moving like a wave along the DNA.

1. RNA polymerase unwinds/"unzips" the DNA by breaking the hydrogen bonds between complimentary nucleotides.
2. RNA nucleotides are paired with complementary DNA bases.
3. RNA sugar-phosphate backbone forms with assistance from RNA polymerase.
4. Hydrogen bonds of the untwisted RNA+DNA helix break, freeing the newly synthesized RNA strand.
5. If the cell has a nucleus, the RNA is further processed and then moves through the small nuclear pores to the cytoplasm.

Transcription is the first step leading to gene expression. The stretch of DNA transcribed into an RNA molecule is called a *transcription unit* and encodes at least one gene. If the gene transcribed encodes a protein, the result of transcription is messenger RNA (mRNA), which will then be used to create that protein via the process of translation. Alternatively, the transcribed gene may encode for either ribosomal RNA (rRNA) or transfer RNA (tRNA), other components of the protein-assembly process, or other ribozymes.

A DNA transcription unit encoding for a protein contains not only the sequence that will eventually be directly translated into the protein (the *coding sequence*) but also *regulatory sequences* that direct and regulate the synthesis of that protein. The regulatory sequence before (upstream from) the coding sequence is called the five prime untranslated region (5'UTR), and the sequence following (downstream from) the coding sequence is called the three prime untranslated region (3'UTR).

Transcription has some proofreading mechanisms, but they are fewer and less effective than the controls for copying DNA; therefore, transcription has a lower copying fidelity than DNA replication.

As in DNA replication, DNA is read from 3' → 5' during transcription. Meanwhile, the complementary RNA is created from the 5' → 3' direction. This means its 5' end is created first in base pairing. Although DNA is arranged as two antiparallel strands in a double helix, only one of the two DNA strands, called the template strand, is used for transcription. This is because RNA is only single-stranded, as opposed to double-stranded DNA. The other DNA strand is called the coding strand, because its sequence is the same as the newly created RNA transcript (except for the substitution of uracil for thymine). The use of only the 3' → 5' strand eliminates the need for the Okazaki fragments seen in DNA replication. Transcription is divided into 5 stages: *pre-initiation*, *initiation*, *promoter clearance*, *elongation* and *termination*.

Ribonucleic Acid

Ribonucleic acid (RNA) is one of the three major macromolecules (along with DNA and proteins) that are essential for all known forms of life. Like DNA, RNA is made up of a long chain of components called nucleotides. Each nucleotide consists of a nucleobase (sometimes called a nitrogenous base), a ribose sugar, and a phosphate group. The sequence of nucleotides allows RNA to encode genetic information. For example, some viruses use RNA instead of DNA as their genetic material, and all organisms use messenger RNA (mRNA) to carry the genetic information that directs the synthesis of proteins.

Like proteins, some RNA molecules play an active role in cells by catalyzing biological reactions, controlling gene expression, or sensing and communicating responses to cellular signals. One of these active processes is protein synthesis, a universal function whereby mRNA molecules direct the assembly of proteins on ribosomes. This process uses transfer RNA (tRNA) molecules to deliver amino acids to the ribosome, where ribosomal RNA (rRNA) links amino acids together to form proteins.

The chemical structure of RNA is very similar to that of DNA, with two differences—(a) RNA contains the sugar ribose while DNA contains the slightly different sugar deoxyribose (a type of ribose that lacks one oxygen atom), and (b) RNA has the nucleobase uracil while DNA contains thymine (uracil and thymine have similar base-pairing properties). Unlike DNA, most RNA molecules are single-stranded. Single-stranded RNA molecules adopt very complex three-dimensional structures, since they are not restricted to the repetitive double-helical form of double-stranded DNA. RNA is made within living cells by RNA polymerases, enzymes that act to copy a DNA or RNA template into a new RNA strand through processes known as transcription or RNA replication, respectively.

Comparison with DNA

RNA and DNA are both nucleic acids, but differ in three main ways. First, unlike DNA, which is, in general, double-stranded, RNA is a single-stranded molecule in many of its biological roles and has a much shorter chain of nucleotides. Second, while DNA contains *deoxyribose*, RNA contains *ribose* (in deoxyribose there is no hydroxyl group attached to the pentose ring in the 2' position). These hydroxyl groups make RNA less stable than DNA because it is more prone to hydrolysis. Third, the complementary base to adenine is not thymine, as it is in DNA, but rather uracil, which is an unmethylated form of thymine. Like DNA, most biologically active RNAs, including mRNA, tRNA, rRNA, snRNAs, and other non-coding RNAs, contain self-complementary sequences that allow parts of the RNA to fold and pair with itself to form double helices. Structural analysis of these RNAs has revealed that they are highly structured. Unlike DNA, their structures do not consist of long double helices but rather collections of short helices packed together into structures akin to proteins. In this fashion, RNAs can achieve chemical catalysis, like enzymes. For instance, determination of the structure of the ribosome—an enzyme that catalyzes peptide bond formation—revealed that its active site is composed entirely of RNA.

Structure

Each nucleotide in RNA contains a ribose sugar, with carbons numbered 1' through 5'. A base is attached to the 1' position, in general, adenine (A), cytosine (C), guanine (G), or uracil (U). Adenine and guanine are purines, cytosine, and uracil are pyrimidines. A phosphate group is attached to the 3' position of one ribose and the 5' position of the next. The phosphate groups have a negative charge each at physiological pH, making RNA a charged molecule (polyanion). The bases may form hydrogen bonds between cytosine and guanine, between adenine and uracil and between guanine and uracil. However, other interactions are possible, such as a group of adenine bases binding to each other in a bulge, or the GNRA tetraloop that has a guanine–adenine base-pair.

An important structural feature of RNA that distinguishes it from DNA is the presence of a hydroxyl group at the 2' position of the ribose sugar. The presence of this functional group causes the helix to adopt the A-form geometry rather than the B-form most commonly observed in DNA. This results in a very deep and narrow major groove and a shallow and wide minor groove. A second consequence of the presence of the 2'-hydroxyl group is that in conformationally flexible regions of an RNA molecule (that is, not involved in formation of a double helix), it can chemically attack the adjacent phosphodiester bond to cleave the backbone.

RNA is transcribed with only four bases (adenine, cytosine, guanine and uracil), but these bases and attached sugars can be modified in numerous ways as the RNAs mature. Pseudouridine (Ø), in which the linkage between uracil and ribose is changed from a C–N bond to a C–C bond, and ribothymidine (T) are found in various places (the most notable ones being in the TØC loop of tRNA). Another notable modified base is hypoxanthine, a deaminated adenine base whose nucleoside is called inosine (I). Inosine plays a key role in the wobble hypothesis of the genetic code.

There are nearly 100 other naturally occurring modified nucleosides, of which pseudouridine and nucleosides with 2'-O-methylribose are the most common. The specific roles of many of these modifications in RNA are not fully understood. However, it is notable that, in ribosomal RNA, many of the post-transcriptional modifications occur in highly functional regions, such as the peptidyl transferase center and the subunit interface, implying that they are important for normal function.

The functional form of single stranded RNA molecules, just like proteins, frequently requires a specific tertiary structure. The scaffold for this structure is provided by secondary structural elements that are hydrogen bonds within the molecule. This leads to several recognizable "domains" of secondary structure like hairpin loops, bulges, and internal loops. Since RNA is charged, metal ions such as Mg^{2+} are needed to stabilise many secondary and tertiary structures.

Synthesis

Synthesis of RNA is usually catalyzed by an enzyme—RNA polymerase—using DNA as a template, a process known as transcription. Initiation of transcription begins with the binding of the enzyme to a promoter sequence in the DNA (usually found "upstream" of a gene). The DNA double helix is unwound by the helicase activity of the enzyme. The enzyme then progresses along the template strand in the 3' to 5' direction, synthesizing a complementary RNA molecule with elongation occurring in the 5' to 3' direction. The DNA sequence also dictates where termination of RNA synthesis will occur. RNAs are often modified by enzymes after transcription. For example, a poly(A) tail and a 5' cap are added to eukaryotic pre-mRNA and introns are removed by the spliceosome. There are also a number of RNA-dependent RNA polymerases that use RNA as their template for synthesis of a new strand of RNA. For instance, a number of RNA viruses (such as poliovirus) use this type of enzyme to replicate their genetic material. Also, RNA-dependent RNA polymerase is part of the RNA interference pathway in many organisms.

Chapter 7

Regulation of Gene Expression

Regulation of gene expression refers to the control of the amount and timing of appearance of the functional product of a gene. Control of expression is vital to allow a cell to produce the gene products it needs when it needs them; in turn this gives cells the flexibility to adapt to a variable environment, external signals, damage to the cell, etc. Some simple examples of where gene expression is important are:

- Control of insulin expression so it gives a signal for blood glucose regulation
- X chromosome inactivation in female mammals to prevent an "overdose" of the genes it contains.
- Cyclin expression levels control progression through the eukaryotic cell cycle

More generally gene regulation gives the cell control over all structure and function, and is the basis for cellular differentiation, morphogenesis and the versatility and adaptability of any organism.

Any step of gene expression may be modulated, from the DNA-RNA transcription step to post-translational modification of a protein. The stability of the final gene product, whether it is RNA or protein, also contributes to the expression level of the gene - an unstable product results in a low expression level. In general gene expression is regulated through changes in the number and type of interactions between molecules that collectively influence transcription of DNA and translation of RNA. Numerous terms are used to describe types of genes depending on how they are regulated, these include:

- A constitutive gene is a gene that is transcribed continually compared to a facultative gene which is only transcribed when needed.
- A housekeeping gene is typically a constitutive gene that is transcribed at a relatively constant level. The housekeeping gene's products are typically needed for maintenance of the cell. It is generally assumed that their expression is unaffected by experimental conditions. Examples include actin, GAPDH and ubiquitin.
- A facultative gene is a gene which is only transcribed when needed compared to a constitutive gene.
- An inducible gene is a gene whose expression is either responsive to environmental change or dependent on the position in the cell cycle.

Transcriptional Regulation

Regulation of transcription can be broken down into three main routes of influence; genetic (direct interaction of a control factor with the gene), modulation (interaction of a control factor with the transcription machinery) and epigenetic (non-sequence changes in DNA structure which influence transcription). Direct interaction with DNA is the simplest and the most direct method by which a protein can change transcription levels. Genes often have several protein binding sites around the coding region with the specific function of regulating transcription. There are many classes of regulatory DNA binding sites known as enhancers, insulators, repressors and silencers. The mechanisms for regulating transcription are very varied, from blocking key binding sites on the DNA for RNA polymerase to acting as an activator and promoting transcription by assisting RNA polymerase binding.

The activity of transcription factors is further modulated by intracellular signals causing protein post-translational modification including phosphorylated, acetylated, or glycosylated. These changes influence a transcription factor's ability to bind, directly or indirectly, to promoter DNA, to recruit RNA polymerase, or to favor elongation of a newly synthetized RNA molecule.

The nuclear membrane in eukaryotes allows further regulation of transcription factors by the duration of their presence in the nucleus which is regulated by reversible changes in their structure and by binding of other proteins. Environmental stimuli or endocrine signals may cause modification of regulatory proteins eliciting cascades of intracellular signals, which result in regulation of gene expression. More recently it has become apparent that there is a huge influence of non-DNA-sequence specific effects on translation. These effects are referred to as epigenetic and involve the higher order structure of DNA, non-sequence specific DNA binding proteins and chemical modification of DNA. In general epigenetic effects alter the accessibility of DNA to proteins and so modulate transcription.

In eukaryotes, DNA is organized in form of nucleosomes. Note how the DNA (blue and green) is tightly wrapped around the protein core made of histone octamer (ribbon coils), restricting access to the DNA. From PDB 1KX5. DNA methylation is a widespread mechanism for epigenetic influence on gene expression and is seen in bacteria and eukaryotes and has roles in heritable transcription silencing and transcription regulation. In eukaryotes the structure of chromatin, controlled by the histone code, regulates access to DNA with significant impacts on the expression of genes in euchromatin and heterochromatin areas.

Regulated Stages of Gene Expression

Any step of gene expression may be modulated, from the DNA-RNA transcription step to post-translational modification of a protein. The following is a list of stages where gene expression is regulated, the most extensively utilised point is Transcription Initiation:

- Chromatin domains
- Transcription
- Post-transcriptional modification
- RNA transport
- Translation
- mRNA degradation

Modification of DNA

In eukaryotes, the accessibility of large regions of DNA can depend on its chromatin structure, which can be altered as a result of histone modifications directed by DNA methylation, ncRNA, or DNA-binding protein. Hence these modifications may up or down regulate the expression of gene. Certain of these modifications that regulate gene expression are inheritable and are referred to as epigenetic regulation.

Chemical

Methylation of DNA is a common method of gene silencing. DNA is typically methylated by methyltransferase enzymes on cytosine nucleotides in a CpG dinucleotide sequence (also called "CpG islands" when densely clustered). Analysis of the pattern of methylation in a given region of DNA (which can be a promoter) can be achieved through a method called bisulfite mapping. Methylated cytosine residues are unchanged by the treatment, whereas unmethylated ones are changed to uracil. The differences are analyzed by DNA sequencing or by methods developed to quantify SNPs, such as Pyrosequencing (Biotage) or MassArray (Sequenom), measuring the relative amounts of C/T at the CG dinucleotide. Abnormal methylation patterns are thought to be involved in oncogenesis.

Structural

Transcription of DNA is dictated by its structure. In general, the density of its packing is indicative of the frequency of transcription. Octameric protein complexes called nucleosomes are responsible for the amount of supercoiling of DNA, and these complexes can be temporarily modified by processes such as phosphorylation or more permanently modified by processes such as methylation. Such modifications are considered to be responsible for more or less permanent changes in gene expression levels. Histone acetylation is also an important process in transcription. Histone acetyltransferase enzymes (HATs) such as CREB-binding protein also dissociate the DNA from the histone complex, allowing transcription to proceed.

Often, DNA methylation and histone deacetylation work together in gene silencing. The combination of the two seems to be a signal for DNA to be packed more densely, lowering gene expression.

Regulation of Transcription

Regulation of transcription controls when transcription occurs and how much RNA is created. Transcription of a gene by RNA polymerase can be regulated by at least five mechanisms:

- Specificity factors alter the specificity of RNA polymerase for a given promoter or set of promoters, making it more or less likely to bind to them (i.e., sigma factors used in prokaryotic transcription).
- Repressors bind to non-coding sequences on the DNA strand that are close to or overlapping the promoter region, impeding RNA polymerase's progress along the strand, thus impeding the expression of the gene.
- General transcription factors position RNA polymerase at the start of a protein-coding sequence and then release the polymerase to transcribe the mRNA.
- Activators enhance the interaction between RNA polymerase and a particular promoter, encouraging the expression of the gene. Activators do this by increasing the attraction of RNA polymerase for the promoter, through interactions with subunits of the RNA polymerase or indirectly by changing the structure of the DNA.
- Enhancers are sites on the DNA helix that are bound to by activators in order to loop the DNA bringing a specific promoter to the initiation complex. Enhancers are much more common in eukaryote than prokaryotes, where only a few examples exist (to date).

Post-transcriptional Regulation

After the DNA is transcribed and mRNA is formed, there must be some sort of regulation on how much the mRNA is translated into proteins. Cells do this by modulating the capping, splicing, addition of a Poly(A) Tail, the sequence-specific nuclear export rates, and, in several contexts, sequestration of the RNA transcript. These processes occur in eukaryotes but not in prokaryotes. This modulation is a result of a protein or transcript that, in turn, is regulated and may have an affinity for certain sequences.

Regulation of Translation

The translation of mRNA can also be controlled by a number of mechanisms, mostly at the level of initiation. Recruitment of the small ribosomal subunit can indeed be modulated by mRNA secondary structure, antisense RNA binding, or protein binding. In both prokaryotes and eukaryotes, a large number of RNA binding proteins exist, which often are directed to their target sequence by the secondary structure of the transcript, which may change depending on certain conditions, such as temperature or presence of a ligand (aptamer). Some transcripts act as ribozymes and self-regulate their expression.

Examples of Gene Regulation

- Enzyme induction is a process in which a molecule (e.g., a drug) induces (i.e., initiates or enhances) the expression of an enzyme.
- The induction of heat shock proteins in the fruit fly Drosophila melanogaster.
- The Lac operon is an interesting example of how gene expression can be regulated.
- Viruses, despite having only a few genes, possess mechanisms to regulate their gene expression, typically into an early and late phase, using collinear systems regulated by anti-terminators (lambda phage) or splicing modulators (HIV).

Developmental Biology

A large number of studied regulatory systems come from developmental biology. Examples include:

- The colinearity of the Hox gene cluster with their nested antero-posterior patterning
- It has been speculated that pattern generation of the hand (digits - interdigits) The gradient of Sonic hedgehog (secreted inducing factor) from the zone of polarizing activity in the limb, which creates a gradient of active Gli3, which activates Gremlin, which inhibits BMPs also secreted in the limb, resulting in the formation of an alternating pattern of activity as a result of this reaction-diffusion system.

- Somitogenesis is the creation of segments (somites) from a uniform tissue (Pre-somitic Mesoderm, PSM). They are formed sequentially from anterior to posterior. This is achieved in amniotes possibly by means of two opposing gradients, Retinoic acid in the anterior (wavefront) and Wnt and Fgf in the posterior, coupled to an oscillating pattern (segmentation clock) composed of FGF + Notch and Wnt in antiphase.
- Sex determination in the soma of a Drosophila requires the sensing of the ratio of autosomal genes to sex chromosome-encoded genes, which results in the production of sexless splicing factor in females, resulting in the female isoform of doublesex.

Morphogen

A morphogen is a substance governing the pattern of tissue development, and the positions of the various specialized cell types within a tissue. It spreads from a localized source and forms a concentration gradient across a developing tissue . In developmental biology a morphogen is rigorously used to mean a signaling molecule that acts directly on cells (not through serial induction) to produce specific cellular responses dependent on morphogen concentration.

Well-known morphogens include: Decapentaplegic / Transforming growth factor beta, Hedgehog / Sonic Hedgehog, Wingless / Wnt, Epidermal growth factor, Fibroblast growth factor, and Retinoic acid. Morphogens are defined conceptually, not chemically , so simple compounds such as retinoic acid (the active metabolite of retinol or vitamin A) may also act as morphogens. During early development, morphogen gradients generate different cell types in distinct spatial order. The morphogen provides spatial information by forming a concentration gradient that subdivides a field of cells by inducing or maintaining the expression of different target genes at distinct concentration thresholds. Thus, cells far from the source of the morphogen will receive low levels of morphogen and express only low-threshold target genes. In contrast, cells close to the source of morphogen will receive high levels of morphogen and will express both low- and high-threshold target genes.

Distinct cell types emerge as a consequence of the different combination of target gene expression. In this way, the field of cells is subdivided into different types according to their position relative to the source of the morphogen. This is a general mechanism by which cell type diversity can be generated in animal development. Some of the earliest and best-studied morphogens are transcription factors that diffuse within early Drosophila melanogaster (fruit fly) embryos. However, most morphogens are secreted proteins that signal between cells.

Fruit Fly

Drosophila melanogaster has an unusual developmental system, in which the first thirteen cell divisions of the embryo occur within a syncytium prior to cellularization. Essentially the embryo remains a single cell with over 8000 nuclei evenly spaced near the membrane until the fourteenth cell division, when independent membranes furrow between the nuclei, separating them into independent cells. As a result, in fly embryos transcription factors such as Bicoid or Hunchback can act as morphogens because they can freely diffuse between nuclei to produce smooth gradients of concentration without relying on specialized intercellular signalling mechanisms. Although there is some evidence that homeobox transcription factors similar to these can pass directly through cell membranes, this mechanism is not believed to contribute greatly to morphogenesis in cellularized systems.

In most developmental systems, such as human embryos or later Drosophila development, syncytia occur only rarely (such as in skeletal muscle), and morphogens are generally secreted signalling proteins. These proteins bind to the extracellular domains of transmembrane receptor proteins, which use an elaborate process of signal transduction to communicate the level of morphogen to the nucleus. The nuclear targets of signal transduction pathways are usually transcription factors, whose activity is regulated in a manner that reflects the level of morphogen received at the cell surface. Thus, secreted morphogens act to generate gradients of transcription factor activity just like those that are generated in the syncitial Drosophila embryo.

Discrete target genes respond to different thresholds of morphogen activity. The expression of target genes is controlled by segments of DNA called 'enhancers' to which transcription factors bind directly. Once bound, the transcription factor then stimulates or inhibits the transcription of the gene and thus controls the level of expression of the gene product (usually a protein). 'Low-threshold' target genes require only low levels of morphogen activity to be regulated and feature enhancers that contain many high-affinity binding sites for the transcription factor. 'High-threshold' target genes have relatively fewer binding sites or low-affinity binding sites that require much greater levels of transcription factor activity to be regulated.

Thus, the general mechanism by which morphogens subdivide tissues into patterns of distinct cell types is well understood. However, morphogens often have additional activities such as controlling the growth of the tissue or orienting the polarity of cells within it (for example, the hairs on your forearm point in one direction). These activities of morphogens are much less understood and are the subject of current research efforts in the field of developmental biology.

Discovery

The morphogen idea has a long history in developmental biology, dating back to the work of the pioneering Drosophila geneticist, Thomas Hunt Morgan, in the early 20th century. However, it was Lewis Wolpert who refined the morphogen concept in the 1960s with his famous French flag model which described how morphogen could subdivide a tissue into domains of different target gene expression (corresponding to the colours of the French flag). This model was championed by the leading Drosophila biologist, Peter Lawrence. Christiane Nusslein-Volhard identified the first morphogen, Bicoid, one of the transcription factors present in a gradient in the Drosophila syncitial embryo. Two labs, that of Gary Struhl and that of Stephen Cohen, then demonstrated that a secreted signalling protein, Decapentaplegic (the Drosophila homologue of Transforming Growth Factor Beta), acted as a morphogen during later stages of Drosophila development.

Subsequent studies of the development of many different animals has confirmed the widespread importance of morphogens in governing animal development.

Gene Regulatory Network

A gene regulatory network or genetic regulatory network (GRN) is a collection of DNA segments in a cell which interact with each other indirectly (through their RNA and protein expression products) and with other substances in the cell, thereby governing the rates at which genes in the network are transcribed into mRNA.

In general, each mRNA molecule goes on to make a specific protein (or set of proteins). In some cases this protein will be structural, and will accumulate at the cell membrane or within the cell to give it particular structural properties. In other cases the protein will be an enzyme, i.e., a micro-machine that catalyses a certain reaction, such as the breakdown of a food source or toxin. Some proteins though serve only to activate other genes, and these are the transcription factors that are the main players in regulatory networks or cascades. By binding to the promoter region at the start of other genes they turn them on, initiating the production of another protein, and so on. Some transcription factors are inhibitory. In single-celled organisms, regulatory networks respond to the external environment, optimising the cell at a given time for survival in this environment. Thus a yeast cell, finding itself in a sugar solution, will turn on genes to make enzymes that process the sugar to alcohol. This process, which we associate with wine-making, is how the yeast cell makes its living, gaining energy to multiply, which under normal circumstances would enhance its survival prospects.

In multicellular animals the same principle has been put in the service of gene cascades that control body-shape. Each time a cell divides, two cells result which, although they contain the same genome in full, can differ in which genes are turned on and making proteins. Sometimes a 'self-sustaining feedback loop' ensures that a cell maintains its identity and passes it on.

Less understood is the mechanism of epigenetics by which chromatin modification may provide cellular memory by blocking or allowing transcription. A major feature of multicellular animals is the use of morphogen gradients, which in effect provide a positioning system that tells a cell where in the body it is, and hence what sort of cell to become. A gene that is turned on in one cell may make a product that leaves the cell and diffuses through adjacent cells, entering them and turning on genes only when it is present above a certain threshold level.

These cells are thus induced into a new fate, and may even generate other morphogens that signal back to the original cell. Over longer distances morphogens may use the active process of signal transduction. Such signalling controls embryogenesis, the building of a body plan from scratch through a series of sequential steps. They also control maintain adult bodies through feedback processes, and the loss of such feedback because of a mutation can be responsible for the cell proliferation that is seen in cancer. In parallel with this process of building structure, the gene cascade turns on genes that make structural proteins that give each cell the physical properties it needs. It has been suggested that, because biological molecular interactions are intrinsically stochastic, gene networks are the result of cellular processes and not their cause (i.e. Cellular Darwinism). However, recent experimental evidence has favored the attractor view of cell fates.

At one level, biological cells can be thought of as "partially-mixed bags" of biological chemicals – in the discussion of gene regulatory networks, these chemicals are mostly the mRNAs and proteins that arise from gene expression. These mRNA and proteins interact with each other with various degrees of specificity. Some diffuse around the cell. Others are bound to cell membranes, interacting with molecules in the environment. Still others pass through cell membranes and mediate long range signals to other cells in a multi-cellular organism. These molecules and their interactions comprise a gene regulatory network. A typical gene regulatory network looks something like this: The nodes of this network are proteins, their corresponding mRNAs, and protein/protein complexes. Nodes that are depicted as lying along vertical lines are associated with the cell/environment interfaces, while the others are free-floating and diffusible. Implied are genes, the DNA sequences which are transcribed into the mRNAs that translate into proteins.

Edges between nodes represent individual molecular reactions, the protein/protein and protein/mRNA interactions through which the products of one gene affect those of another, though the lack of experimentally obtained information often implies that some reactions are not modeled at such a fine level of detail. These interactions can be inductive (the arrowheads), with an increase in the concentration of one leading to an increase in the other, or inhibitory (the filled circles), with an increase in one leading to a decrease in the other. A series of edges indicates a chain of such dependences, with cycles corresponding to feedback loops. The network structure is an abstraction of the system's chemical dynamics, describing the manifold ways in which one substance affects all the others to which it is connected. In practice, such GRNs are inferred from the biological literature on a given system and represent a distillation of the collective knowledge about a set of related biochemical reactions.

Genes can be viewed as nodes in the network, with input being proteins such as transcription factors, and outputs being the level of gene expression. The node itself can also be viewed as a function which can be obtained by combining basic functions upon the inputs (in the Boolean network described below these are Boolean functions, typically AND, OR, and NOT). These functions have been interpreted as performing a kind of information processing within the cell, which determines cellular behavior. The basic drivers within cells are concentrations of some proteins, which determine both spatial (location within the cell or tissue) and temporal (cell cycle or developmental stage) coordinates of the cell, as a kind of "cellular memory".

The gene networks are only beginning to be understood, and it is a next step for biology to attempt to deduce the functions for each gene "node", to help understand the behavior of the system in increasing levels of complexity, from gene to signaling pathway, cell or tissue level. Mathematical models of GRNs have been developed to capture the behavior of the system being modeled, and in some cases generate predictions corresponding with experimental observations. In some other cases, models have proven to make accurate novel predictions, which can be tested experimentally, thus suggesting new approaches to explore in an experiment that sometimes wouldn't be considered in the design of the protocol of an experimental laboratory. The most common modeling technique involves the use of coupled ordinary differential equations (ODEs). Several other promising modeling techniques have been used, including Boolean networks, Petri nets, Bayesian networks, graphical

Gaussian models, Stochastic, and Process Calculi. Conversely, techniques have been proposed for generating models of GRNs that best explain a set of time series observations.

Modelling

Coupled ODEs

For an example of modelling of the cell cycle with ODEs, see cellular model. It is common to model such a network with a set of coupled ordinary differential equations (ODEs) or stochastic ODEs, describing the reaction kinetics of the constituent parts. Suppose that our regulatory network has N nodes, and let $S_1(t), S_2(t), ..., S_N(t)$ represent the concentrations of the N corresponding substances at time t. Then the temporal evolution of the system can be described approximately by

$$\frac{dS_j}{dt} = f_j(S_1, S_2, ..., S_N)$$

where the functions f_j express the dependence of S_j on the concentrations of other substances present in the cell. The functions f_j are ultimately derived from basic principles of chemical kinetics or simple expressions derived from these e.g. Michaelis-Menten enzymatic kinetics. Hence, the functional forms of the f_j are usually chosen as low-order polynomials or Hill functions that serve as an ansatz for the real molecular dynamics. Such models are then studied using the mathematics of nonlinear dynamics. System-specific information, like reaction rate constants and sensitivities, are encoded as constant parameters.

By solving for the fixed point of the system:

$$\frac{dS_j}{dt} = 0$$

for all j, one obtains (possibly several) concentration profiles of proteins and mRNAs that are theoretically sustainable (though not necessarily stable). Steady states of kinetic equations thus correspond to potential cell types, and oscillatory solutions to the above equation to naturally cyclic cell types. Mathematical stability of these attractors can usually be characterized by the sign of higher derivatives at critical points, and then correspond to biochemical stability of the concentration profile. Critical points and bifurcations in the equations correspond to critical cell states in which small state or parameter perturbations could switch the system between one of several stable differentiation fates. Trajectories correspond to the unfolding of biological pathways and transients of the equations to short-term biological events. For a more mathematical discussion, see the articles on nonlinearity, dynamical systems, bifurcation theory, and chaos theory.

Boolean Network

The following example illustrates how a Boolean network can model a GRN together with its gene products (the outputs) and the substances from the environment that affect it (the inputs). Stuart Kauffman was amongst the first biologists to use the metaphor of Boolean networks to model genetic regulatory networks.

1. Each gene, each input, and each output is represented by a node in a directed graph in which there is an arrow from one node to another if and only if there is a causal link between the two nodes.
2. Each node in the graph can be in one of two states: on or off.
3. For a gene, "on" corresponds to the gene being expressed; for inputs and outputs, "on" corresponds to the substance being present.
4. Time is viewed as proceeding in discrete steps. At each step, the new state of a node is a Boolean function of the prior states of the nodes with arrows pointing towards it.

The validity of the model can be tested by comparing simulation results with time series observations.

Continuous Networks

Continuous network models of GRNs are an extension of the boolean networks described above. Nodes still represent genes and connections between them regulatory influences on gene expression. Genes in biological systems display a continuous range of activity levels and it has been argued that using a continuous representation captures several properties of gene regulatory networks not present in the Boolean model. Formally most of these approaches are similar to an artificial neural network, as inputs to a node are summed up and the result serves as input to a sigmoid function, e.g., but proteins do often control gene expression in a synergistic, i.e. non-linear, way. However there is now a continuous network model that allows grouping of inputs to a node thus realizing another level of regulation. This model is formally closer to a higher order recurrent neural network. The same model has also been used to mimic the evolution of cellular differentiation and even multicellular morphogenesis.

Stochastic Gene Networks

Recent experimental results have demonstrated that gene expression is a stochastic process. Thus, many authors are now using the stochastic formalism, after the work by. Works on single gene expression and small synthetic genetic networks, such as the genetic toggle switch of Tim Gardner and Jim Collins, provided additional experimental data on the phenotypic variability and the stochastic nature of gene expression. The first versions of stochastic models of gene expression involved only instantaneous reactions and were driven by the Gillespie algorithm.

Since some processes, such as gene transcription, involve many reactions and could not be correctly modeled as an instantaneous reaction in a single step, it was proposed to model these reactions as single step multiple delayed reactions in order to account for the time it takes for the entire process to be complete. From here, a set of reactions were proposed that allow generating GRNs. These are then simulated using a modified version of the Gillespie algorithm, that can simulate multiple time delayed reactions (chemical reactions where each of the products is provided a time delay that determines when will it be released in the system as a "finished product").

For example, basic transcription of a gene can be represented by the following single-step reaction (RNAP is the RNA polymerase, RBS is the RNA ribosome binding site, and Pro $_i$ is the promoter region of gene i):

$$RNAP + \mathrm{Pr}\,o_i \xrightarrow{k_i,bas} \mathrm{Pr}\,o_i\left(\tau_i^1\right) + RBS_i\left(\tau_i^1\right) + RNAp\left(\tau_I^2\right)$$

A recent work proposed a simulator (SGNSim, Stochastic Gene Networks Simulator), that can model GRNs where transcription and translation are modeled as multiple time delayed events and its dynamics is driven by a stochastic simulation algorithm (SSA) able to deal with multiple time delayed events. The time delays can be drawn from several distributions and the reaction rates from complex functions or from physical parameters. SGNSim can generate ensembles of GRNs within a set of user-defined parameters, such as topology. It can also be used to model specific GRNs and systems of chemical reactions. Genetic perturbations such as gene deletions, gene over-expression, insertions, frame shift mutations can also be modeled as well.

The GRN is created from a graph with the desired topology, imposing in-degree and out-degree distributions. Gene promoter activities are affected by other genes expression products that act as inputs, in the form of monomers or combined into multimers and set as direct or indirect. Next, each direct input is assigned to an operator site and different transcription factors can be allowed, or not, to compete for the same operator site, while indirect inputs are given a target. Finally, a function is assigned to each gene, defining the gene's response to a combination of transcription factors (promoter state). The transfer functions (that is, how genes respond to a combination of inputs) can be assigned to each combination of promoter states as desired. In other recent work, multiscale models of gene regulatory networks have been developed that focus on synthetic biology applications. Simulations have been used that model all biomolecular interactions in transcription, translation, regulation, and induction of gene regulatory networks, guiding the design of synthetic systems.

Prediction

Other work has focused on predicting the gene expression levels in a gene regulatory network. The approaches used to model gene regulatory networks have been constrained to be interpretable and, as a result, are generally simplified versions of the network. For example, Boolean networks have been used due to their simplicity and ability to handle noisy data but lose data information by having a binary representation of the genes.

Also, artificial neural networks omit using a hidden layer so that they can be interpreted, losing the ability to model higher order correlations in the data. Using a model that is not constrained to be interpretable, a more accurate model can be produced. Being able to predict gene expressions more accurately provides a way to explore how drugs affect a system of genes as well as for finding which genes are interrelated in a process. This has been encouraged by the DREAM competition which promotes a competition for the best prediction algorithms. Some other recent work has used artificial neural networks with a hidden layer.

Network Connectivity

Empirical data indicate that biological gene networks are sparsely connected, and that the average number of upstream-regulators per gene is less than two. Theoretical results show that selection for robust gene networks will favor minimally complex, more sparsely connected, networks. These results suggest that a sparse, minimally connected, genetic architecture may be a fundamental design constraint shaping the evolution of gene network complexity.

Body Plan

A body plan is the blueprint for the way the body of an organism is laid out. An organism's symmetry, its number of body segments and number of limbs are all aspects of its body plan. One of the key issues of developmental biology is the evolution of body plans as different as those of a starfish, a fern, or a mammal, from a common biological heritage, and in particular how radical changes in body plans have occurred over geological time. The body plan is a key feature of an organism's morphology and, since the discovery of DNA, developmental biologists have been able to learn a lot about how genes control the development of structural features through a cascade of processes in which key genes produce morphogens, chemicals that diffuse through the body to produce a gradient that acts as a position indicator for cells, turning on other genes, some of which in turn produce other morphogens.

A key discovery was the existence of groups of homeobox genes which are responsible for laying down the basic body plan in organisms. The homeobox genes are remarkably conserved between species as diverse as the fruitfly and man, the basic segmented pattern of the worm or fruitfly being the origin of the segmented spine in man. The field of evolutionary developmental biology, which studies the genetics of morphology in detail is now a rapidly expanding one , with many of the developmental genetic cascades, particularly in the fruitfly drosophila, now catalogued in considerable detail. Body plan is the basis for phylum, and there are 35 different basic animal body plans, corresponding to different phyla.

Origin

The evolution of body plans became inevitable with the emergence of differentiated multicellular life in the Ediacaran Era, over 600 million years ago. The most basic and successful structure, for free-moving organisms, is the "pipe" or alimentary canal. This is common even to organisms as diverse as humans and earthworms. It is essentially a passage having a mouth at one end, and a cloaca or anus at the other. The simple process of nutrient capture, digestion, and waste disposal is fundamental to the body plan of advanced, free-moving animals. Vertebra, limbs, even brains are supplementary to the pipe. Natural selection has spun off an enormous range of variations on this basic theme, but the pipe model itself remains. The basic symmetry and organization of this body plan apparently gave an ancient organism an enormous advantage at survival and reproduction, and it has been preserved in most animals ever since. The Cambrian explosion refers to the massive increase in different body plans that took place around 530 million years ago. Fossils from this era show all sorts of odd shapes, many quite unlike anything found today. At that time it was possible for organisms to survive and make a living even though they were unrefined and unlikely, because predation had yet to evolve, along with arms races that would optimise and streamline them to occupy a particular ecological niche.

Bauplan

Bauplan (German for building plan, blueprint; plural: baupläne or bauplaene) is a closely related term in biology referring to the common new and original (homologous) properties of the members of a systematic group (taxon). It is not necessary that a bauplan precisely describes any one particular species of that group. The concept of bauplan is employed in the studies of morphology, taxonomy, comparative physiology, evolutionary physiology, and, most notably, phylogenetics and evolution. Before the advent of genetic sequencing, the analysis of the bauplan of fossils was the primary method to determine hypothetical relationships and lineages of species, both living and extinct. The idea is that species that are closely related share more common properties, hence a more detailed bauplan. Small differences of bauplan are indicative of species belonging to a parent, child or sibling taxon.

Genetic Basis

Similarities and differences in adult shape and form, as well as the developmental pattern of embryos, provide the framework for modern taxonomic classification. These comparisons are the basis of phylogenetic systematics. Embryonic development is relatively consistent among animals with similar body plans, although similar larval forms may give rise to very different adults in some groups. The timing, pattern, and scale of developmental events determine the shape of an organism, and closely related groups are more likely to share structural and developmental similarities than those that are more distantly related. Homologous structures and developmental stages—those that are similar among related groups because they are inherited from a common ancestor—are the basis of modern biological classification.

The most commonly seen body plan amongst vertebrates is that of the tetrapod, which include all mammals, birds, amphibians and reptiles. Some animal groups, such as the cetaceans, bats and most birds have been modified (e.g. front limbs become wings or flippers) but nevertheless, they are still tetrapods. The invertebrates employ a much more diverse array of body plans, such as seen in insects (six legs, three body parts and an exoskeleton), cephalopods (no skeleton, hydrostatically stiffened tentacles, primary propulsion by squeezing water out of a mantle cavity), echinoderms (fivefold radial symmetry, external skeleton, movement by hydrostatically operated tube feet) and various phyla of "worms" (tube-shaped, movement by expanding and contracting parts of the body). The most varied collection of body forms known is found in the Burgess Shale, where fossils from a Cambrian sea show a tremendous variety of body forms that came to rise (only to later fall extinct) during the Cambrian explosion.

Fictional

One common theme in science fiction is the appearance of extraterrestrial beings, descriptions of which have ranged from being simple variants on human anatomy to beings with body plans wildly different from any found on Earth. The field of exobiology attempts to bring these and similar speculations into the realm of serious scientific investigation.

Cis-regulatory module

Cis-regulatory module (CRM) is a stretch of DNA, usually 100-1000 DNA base pairs in length, where a number of transcription factors can bind and regulate expression of nearby genes. One cis-regulatory element can regulate several genes, and conversely, one gene can have several cis-regulatory modules. Cis-regulatory modules are one of several types of functional regulatory elements. Regulatory elements are binding sites for transcription factors, which are involved in gene regulation. Cis-regulatory modules perform a large amount of developmental information processing. Cis-regulatory modules are non-random clusters at their specified target site that contain transcription factor binding sites.

They are labeled as cis because they are typically located on the same DNA as the genes they control as opposed to trans, which refers to effects on genes not located on the same strand or farther away, such as transcription factors. The original definition presented cis-regulatory modules as enhancers of cis-acting DNA, which increased the rate of transcription from a linked promoter. However, this definition has changed to define cis-regulatory modules as a DNA sequence with transcription factor binding sites which are clustered into modular structures, including -but not limited to- locus control regions, promoters, enhancers, silencers, boundary control elements and other modulators.

Cis-regulatory modules can be divided into classes. Enhancers regulate gene expression positively. Insulators work indirectly by interacting with other nearby cis-regulatory modules. Silencers act by silencing genes. Cis-regulatory modules carry out their function by integrating the active transcription factors and the associated co-factors at a specific time and place in the cell, this information is read and an output is given.

Gene-regulation function of a cis-regulatory module

The design of cis-regulatory modules is such that transcription factors and epigenetic modifications serve as inputs, and the output of the module is the command given to the transcription machinery, which in turn determines the rate of gene transcription or whether it is turned on or off. There are two types of transcription factor inputs: those that determine when the target gene is to be expressed and those that serve as functional drivers, which come into play only during specific situations during development.

These inputs can come from different time points, can represent different signal ligands, or can come from different domains or lineages of cells. However, a lot still remains unknown. Additionally, the regulation of chromatin structure and nuclear organization also play a role in determining and controlling the function of cis-regulatory modules. Thus gene-regulation functions (GRF) provide a unique characteristic of a cis-regulatory module (CRM), relating the concentrations of transcription factors (input) to the promoter activities (output). The challenge is to predict GRFs. This challenge still remains unsolved. In general, gene-regulation functions do not use Boolean logic, although in some cases the approximation of the Boolean logic is still very useful.

The Boolean Logic Assumption

Within the assumption of the Boolean logic, four principles guiding the operation of these cis-regulatory modules are : 1) The design of the cis-regulatory module determines the regulatory function. 2) In relation to development, these modules can generate both positive and negative outputs. 3) The output of each module is a product of the various operations performed on it. Common operations: "OR" logic gate- This design indicates that in an output will be given when either input is given . "AND" logic gate– In this design two different regulatory factors are necessary to make sure that a positive output results.

"Toggle Switches"- This design occurs when the signal ligand is absent while the transcription factor is present; this transcription factor ends up acting as a dominant repressor. However, once the signal ligand is present the transcription factor's role as repressor is eliminated and transcription can occur. Other Boolean logic operations can occur as well, such as sequence specific transcriptional repressors, which when they bind to the cis-regulatory module lead to an output of zero. Additionally, besides influence from the different logic operations, the output of a "cis"-regulatory module will also be influenced by prior events. 4) Cis-regulatory modules must interact with other regulatory elements. For the most part, even with the presence of functional overlap between cis-regulatory modules of a gene, the modules' inputs and outputs tend to not be the same.

Identification and Computational Prediction of CRMs

Besides experimentally determining CRMs, there are various bioinformatics algorithms for predicting them. Most algorithms try to search for significant combinations of transcription factor binding sites (DNA binding sites) in promoter sequences of co-expressed genes. More advanced methods combine the search for significant motifs with correlation in gene expression datasets between transcription factors and target genes. Both methods have been implemented, for example, in the ModuleMaster. Other programs created for the identification and prediction of "cis-regulatory modules include:

Stubb uses hidden Markov models to identify statistically significant clusters of transcription factor combinations. It also uses a second related genome to improve the prediction accuracy of the model. Bayesian Networks use an algorithm that combines site predictions and tissue-specific expression data for transcription factors and target genes of interest. This model also uses regression trees to depict the relationship between the identified cis-regulatory module and the possible binding set of transcription factors.

CRÈME examine clusters of target sites for transcription factors of interest. This program uses a database of confirmed transcription factor binding sites that were annotated across the human genome. A search algorithm is applied to the data set to identify possible combinations of transcription factors, which have binding sites that are close to the promoter of the gene set of interest. The possible cis-regulatory modules are then statistically analyzed and the significant combinations are graphically represented. Active cis-regulatory modules in a genomic sequence have been difficult to identify. Problems in identification arise because often scientists find themselves with a small set of known transcription factors, so it makes it harder to identify statistically significant clusters of transcription factor binding sites. Additionally, high costs limit the use of large whole genome tiling arrays.

Classification of Cis-regulatory Modules

Cis-regulatory modules can be characterized by the information processing that they encode and the organization of their transcription factor binding sites. Additionally, cis-regulatory modules are also characterized by the way they affect the probability, proportion, and rate of transcription. Highly cooperative and coordinated cis-regulatory modules are classified as enhanceosomes. The architecture and the arrangement of the transcription factor binding sites are critical because disruption of the arrangement could cancel out the function. Functional flexible cis-regulatory modules are called billboards. Their transcriptional output is the summation effect of the bound transcription factors.

Enhancers affect the probability of a gene being activated, but have little or no effect on rate. Binary response model acts like an on/off switch for transcription. This model will increase or decrease the amount of cells that transcribe a gene, however it does not affect the rate of transcription. Rheostatic response model describes cis-regulatory modules as regulators of the initiation rate of transcription of its associated gene.

Mode of Action

Cis-regulatory modules can regulate their target genes over large distances, several models have been proposed to describe the way that these modules may communicate with their target gene promoter. DNA Scanning Model: In this model the transcription factor and cofactor complex form at the cis-regulatory module and then continues to move along the DNA sequence until it finds the target gene promoter. Looping Model: In this model the transcription factor binds to the cis-regulatory module, which then causes the looping of the DNA sequence and allows for the interaction with the target gene promoter. Facilitated Tracking Model: This model combines parts of the two previous models. The transcription factor-cis-regulatory module complex causes the looping of the DNA sequence slowly towards the target promoter and forms a stable looped configuration.

Cis-regulatory Module in Gene Regulatory Network

The function of a gene regulatory network depends on the architecture of the nodes, whose function is dependent on the multiple cis-regulatory modules. The layout of cis-regulatory modules can provide enough information to generate spatial and temporal patterns of gene expression. During development each domain, where each domain represents a different spatial regions of the embryo, of gene expression will be under the control of different cis-regulatory module(s). The design of regulatory modules help in producing feedback, feed forward, and cross-regulatory loops.

Genenetwork (database)

GeneNetwork is a database and open source bioinformatics software resource for systems genetics. This resource is used to study gene regulatory networks that link DNA sequence variants to corresponding differences in gene and protein expression and to differences in traits such as health and disease risk. Data sets in GeneNetwork are typically made up of large collections of genotypes (e.g., SNPs) and phenotypes that are obtained from groups of related individuals, including human families, experimental crosses of strains of mice and rats, and organisms as diverse as Drosophila melanogaster, Arabidopsis thaliana, and barley. The inclusion of genotypes for all individuals makes it practical to carry out web-based gene mapping to discover those regions of the genome that contribute to differences in gene expression, cell function, anatomy, physiology, and behavior among individuals.

GeneNetwork was created at the University of Tennessee Health Science Center, Memphis USA in 2000-2001. It was initially developed as a web-adapted version of Kenneth F. Manly's Map Manager QT and QTX programs and was called WebQTL. Gene mapping data were incorporated for several mouse recombinant inbred strains. By early 2003, the first large Affymetrix gene expression data sets (whole mouse brain mRNA and hematopoietic stem cells) were incorporated and the system was renamed. GeneNetwork is now developed by an international group of developers and has mirror and development sites in Europe, Asia, and Australia. The production service is hosted on the Amazon Elastic Compute Cloud.

Organization and Use

GeneNetwork consists of two major components:

- Massive collections of genetic, genomic, and phenotype data for large families
- Sophisticated statistical analysis and gene mapping software that enable analysis of regulatory networks and genotype-to-phenotype relations

Four levels of data are usually obtained for each family or population:

1. DNA sequences and genotypes
2. Gene expression values using microarray, RNA-seq, or proteomic methods (molecular phenotypes)
3. Standard phenotypes of the type that are part of a typical medical record (e.g., blood chemistry, body weight)
4. Annotation files and metadata

The combined data types are housed together in a single relational database, but are conceptually organized and divided by species and family. The system is implemented as a LAMP (software bundle) stack. Code and a simplified version of the MySQL database are available at Sourceforge.net/projects/genenetwork. GeneNetwork is primarily used by researchers but has also been adopted successfully for undergraduate courses in genetics (see YouTube example), bioinformatics, physiology, and psychology. Researchers and students typically retrieve sets of genotypes and phenotypes from one or more families and use built-in statistical and mapping functions to explore relations among variables and to assemble networks of associations. Key steps include the analysis of these factors:

1. The range of variation of traits
2. Covariation among traits (scatterplots and correlations)
3. Architecture of larger networks of traits

4. Quantitative trait locus mapping and causal models of the linkage between sequence differences and phenotype differences

Data Sources

Massive expression data sets are submitted by researchers directly or are extracted from repositories such as National Center for Biotechnology Information Gene Expression Omnibus. A wide variety of cells and tissues are included—from single cell populations of the immune system, specific tissues (retina, prefrontal cortex), to entire systems (whole brain, lung, muscle, heart, fat, kidney, flower, even whole plant embryos). A typical data set is often based on hundreds of fully genotyped individuals and may also include biological replicates. Genotypes and phenotypes are taken from peer-reviewed papers. GeneNetwork includes annotation files for several RNA profiling platforms (Affymetrix, Illumina, and Agilent). RNA-seq data are also available for BXD recombinant inbred mice. Content and nomenclature are reviewed and edited by curators. Updates on coverage of species, families, tissues and measurement types are available at this site: .

Topics of annotation include the following:

- DNA sequence (SNPs, CNVs, indels)
- transcriptomes (arrays, RNA-seq)
- gene regulatory networks
- phenome

Predicting the Number and Organization of Operons

The number and organization of operons has been studied most critically in E. coli. As a result, predictions can be made based on an organism's genomic sequence. One prediction method uses the intergenic distance between reading frames as a primary predictor of the number of operons in the genome. The separation merely changes the frame and guarantees that the read through is efficient. Longer stretches exist where operons start and stop, often up to 40–50 bases. An alternative method to predict operons is based on finding gene clusters where gene order and orientation is conserved in two or more genomes.

Operon prediction is even more accurate if the functional class of the molecules is considered. Bacteria have clustered their reading frames into units, sequestered by co-involvement in protein complexes, common pathways, or shared substrates and transporters. Thus, accurate prediction would involve all of these data, a difficult task indeed. Pascale Cossart published in 2009 the first full map of an operon, identifying the genetic switches that operate in Listeria under different conditions.

Synexpression

Synexpression is a type of non-random eukaryotic gene organization. Genes in a synexpression group may not be physically linked, but they are involved in the same process and they are coordinately expressed. It is expected that genes that function in the same process be regulated coordinately. Synexpression groups in particular represent genes that are simultaneously up- or down-regulated, often because their gene products are required in stoichiometric amounts or are protein-complex subunits. It is likely that these gene groups share common cis- and trans-acting control elements to achieve coordinate expression. Synexpression groups are determined mainly by analysis of expression profiles compiled by the use of DNA microarrays. The use of this technology helps researchers monitor changes in expression patterns for large numbers of genes in a given experiment. Analysis of DNA microarray expression profiles has led to the discovery of a number of genes that are tightly co-regulated.

One simplified example of a synexpression group is the genes cdc6, cdc3, cdc46, and swi4 in yeast, which are all co-expressed early in the G-1 stage of the cell cycle., These genes share one common cis-regulatory element, called ECB, which serves as a binding site for the MCM1 trans-acting protein. Although these genes are not spatially clustered, co-regulation seems to be achieved via this common cis and trans control mechanism. Most synexpression groups are more complicated than the ECB group in yeast, involving a myriad of cis and trans control elements. The identification of synexpression groups has had an impact on the way some scientists view evolutionary change in higher eukaryotes. Since groups of genes involved in the same biological process often share one or more common control elements, it has been suggested that the differential expression of these synexpression groups in different tissues of organisms can contribute to co-evolution tissues, organs, and appendages.

Today it is commonly believed that it is not primarily the gene products themselves that evolve, but that it is the control networks for groups of genes that contribute most to the evolution of higher eukaryotes. Developmental processes provide an example of how changes in synexpression control networks could have a significant impact on an organism's capacity to evolve and adapt effectively. In animals, it is often beneficial for appendages to co-evolve, and it has been observed that fore- and hind-limbs share expression of Hox genes early in metazoan development. Thus, changes in the regulatory patterns of these genes would effect the development of both the fore- and hind-limbs, facilitating co-evolution.

Systems Biology

Systems biology is a term used to describe a number of trends in bioscience research, and a movement which draws on those trends. Proponents describe systems biology as a biology-based inter-disciplinary study field that focuses on complex interactions in biological systems, claiming that it uses a new perspective (holism instead of reduction). Particularly from year 2000 onwards, the term is used widely in the biosciences, and in a variety of contexts. An often stated ambition of systems biology is the modeling and discovery of emergent properties, properties of a system whose theoretical description is only possible using techniques which fall under the remit of systems biology. These typically involve cell signaling networks, via long-range allostery.

Systems biology can be considered from a number of different aspects:

- As a field of study, particularly, the study of the interactions between the components of biological systems, and how these interactions give rise to the function and behavior of that system (for example, the enzymes and metabolites in a metabolic pathway).
- As a paradigm, usually defined in antithesis to the so-called reductionist paradigm (biological organisation), although fully consistent with the scientific method. The distinction between the two paradigms is referred to in these quotations:

"The reductionist approach has successfully identified most of the components and many of the interactions but, unfortunately, offers no convincing concepts or methods to understand how system properties emerge...the pluralism of causes and effects in biological networks is better addressed by observing, through quantitative measures, multiple components simultaneously and by rigorous data integration with mathematical models" Sauer et al

"Systems biology...is about putting together rather than taking apart, integration rather than reduction. It requires that we develop ways of thinking about integration that are as rigorous as our reductionist programmes, but different....It means changing our philosophy, in the full sense of the term" Denis Noble

- As a series of operational protocols used for performing research, namely a cycle composed of theory, analytic or computational modelling to propose specific testable hypotheses about a biological system, experimental validation, and then using the newly acquired quantitative description of cells or cell processes to refine the computational model or theory. Since the objective is a model of the interactions in a system, the experimental techniques that most suit systems biology are those that are system-wide and attempt to be as complete as possible. Therefore, transcriptomics, metabolomics, proteomics and high-throughput techniques are used to collect quantitative data for the construction and validation of models.
- As the application of dynamical systems theory to molecular biology.
- As a socioscientific phenomenon defined by the strategy of pursuing integration of complex data about the interactions in biological systems from diverse experimental sources using interdisciplinary tools and personnel.

This variety of viewpoints is illustrative of the fact that systems biology refers to a cluster of peripherally overlapping concepts rather than a single well-delineated field. However the term has widespread currency and popularity as of 2007, with chairs and institutes of systems biology proliferating worldwide.

History

Systems biology finds its roots in:

- the quantitative modeling of enzyme kinetics, a discipline that flourished between 1900 and 1970,
- the mathematical modeling of population growth,
- the simulations developed to study neurophysiology, and
- control theory and cybernetics.

One of the theorists who can be seen as one of the precursors of systems biology is Ludwig von Bertalanffy with his general systems theory. One of the first numerical simulations in biology was published in 1952 by the British neurophysiologists and Nobel prize winners Alan Lloyd Hodgkin and Andrew Fielding Huxley, who constructed a mathematical model that explained the action potential propagating along the axon of a neuronal cell. Their model described a cellular function emerging from the interaction between two different molecular components, a potassium and a sodium channel, and can therefore be seen as the beginning of computational systems biology. In 1960, Denis Noble developed the first computer model of the heart pacemaker. The formal study of systems biology, as a distinct discipline, was launched by systems theorist Mihajlo Mesarovic in 1966 with an international symposium at the Case Institute of Technology in Cleveland, Ohio entitled "Systems Theory and Biology".

The 1960s and 1970s saw the development of several approaches to study complex molecular systems, such as the Metabolic Control Analysis and the biochemical systems theory. The successes of molecular biology throughout the 1980s, coupled with a skepticism toward theoretical biology, that then promised more than it achieved, caused the quantitative modelling of biological processes to become a somewhat minor field.

However the birth of functional genomics in the 1990s meant that large quantities of high quality data became available, while the computing power exploded, making more realistic models possible. In 1997, the group of Masaru Tomita published the first quantitative model of the metabolism of a whole (hypothetical) cell.

Around the year 2000, after Institutes of Systems Biology were established in Seattle and Tokyo, systems biology emerged as a movement in its own right, spurred on by the completion of various genome projects, the large increase in data from the omics (e.g. genomics and proteomics) and the accompanying advances in high-throughput experiments and bioinformatics.

Since then, various research institutes dedicated to systems biology have been developed. For example, the NIGMS of NIH established a project grant that is currently supporting over ten Systems Biology Centers in the United States. As of summer 2006, due to a shortage of people in systems biology several doctoral training programs in systems biology have been established in many parts of the world. In that same year, the National Science Foundation (NSF) put forward a grand challenge for systems biology in the 21st century to build a mathematical model of the whole cell. In 2011, V. A. Shiva Ayyadurai and C. Forbes Dewey, Jr. of Department of Biological Engineering at the Massachusetts Institute of Technology created CytoSolve, a method to model the whole cell by dynamically integrating multiple molecular pathway models.

Chapter 8

Clinical Trials

On January 23, 2009, the US Food and Drug Administration gave clearance to Geron Corporation for the initiation of the first clinical trial of an embryonic stem cell-based therapy on humans. The trial will evaluate the drug GRNOPC1, embryonic stem cell-derived oligodendrocyte progenitor cells, on patients with acute spinal cord injury. As of mid 2010 hundreds of phase III clinical trials involving stem cells have been registered.

Stem Cell Use in Animals

Veterinary Applications

Potential Contributions to Veterinary Medicine

Research currently conducted on horses, dogs, and cats can benefit the development of stem-cell treatments in veterinary medicine and can target a wide range of injuries and diseases such as myocardial infarction, stroke, tendon and ligament damage, osteoarthritis, osteochondrosis and muscular dystrophy both in large animals, as well as humans.

While investigation of cell-based therapeutics generally reflects human medical needs, the high degree of frequency and severity of certain injuries in racehorses has put veterinary medicine at the forefront of this novel regenerative approach. Companion animals can serve as clinically relevant models that closely mimic human disease.

Development of Regenerative Treatment Models

Veterinary applications of stem cell therapy as a means of tissue regeneration have been largely shaped by research that began with the use of adult-derived mesenchymal stem cells to treat animals with injuries or defects affecting bone, cartilage, ligaments and/or tendons. Because mesenchymal stem cells can differentiate into the cells that make up bone, cartilage, tendons, and ligaments (as well as muscle, fat, and possibly other tissues), they have been the main type of stem cells studied in the treatment of diseases affecting these tissues.

Mesenchymal stem cells are primarily derived from adipose tissue or bone marrow. Since an elevated immune response following cell transplantation may result in rejection of exogenous cells (except in the case of cells derived from a very closely genetically related individual), mesenchymal stem cells are often derived from the patient prior to injection in a process known as autologous transplantation. Surgical repair of bone fractures in dogs and sheep has demonstrated that engraftment of mesenchymal stem cells derived from a genetically different donor within the same species, termed allogeneic transplantation, does not elicit an immunological response in the recipient animal and can mediate regeneration of bone tissue in major bony fractures and defects.

Stem cells can speed up bone repair in fractures/defects that would normally require extensive grafting, suggesting that mesenchymal stem cell use may provide a useful alternative to conventional grafting techniques. Treating tendon and ligament injuries in horses using stem cells, whether derived from adipose tissue or bone-marrow, has support in the veterinary literature. While further studies are necessary to fully characterize the use of cell-based therapeutics for treatment of bone fractures, stem cells are thought to mediate repair via five primary mechanisms: 1) providing an antiinflammatory effect, 2) homing to damaged tissues and recruiting other cells, such as endothelial progenitor cells, that are necessary for tissue growth, 3) supporting tissue remodeling over scar formation, 4) inhibiting apoptosis, and 5) differentiating into bone, cartilage, tendon, and ligament tissue.

Significance of Stem Cell Microenvironments

The microenvironment into which stem cells are transplanted significantly alters the capacity of engrafted cells for recovery and repair. The microenviroment provides growth factors and other chemical signals that guide appropriate differentiation of transplanted cell populations and direct transplanted cells to sites of trauma or disease. Repair and recovery can then be mediated via three primary mechanisms: 1) formation and/or recruitment of new blood cells to the damaged region; 2) prevention of programed cell death or apoptosis; and 3) suppression of inflammation. To further enrich blood supply to the damaged areas, and consequently promote tissue regeneration, platelet-rich plasma could be used in conjunction with stem cell transplantation. The efficacy of some stem cell populations may also be affected by the method of delivery; for instance, to regenerate bone, stem cells are often introduced in a scaffold where they produce the minerals necessary for generation of functional bone.

Sources of Autologous (Patient-derived) Stem Cells

Autologous stem cells intended for regenerative therapy are generally isolated either from the patient's bone marrow or from adipose tissue. The number of stem cells transplanted into damaged tissue may alter efficacy of treatment.

Accordingly, stem cells derived from bone marrow aspirates, for instance, are cultured in specialized laboratories for expansion to millions of cells. Although adipose-derived tissue also requires processing prior to use, the culturing methodology for adipose-derived stem cells is not as extensive as that for bone marrow-derived cells. While it is thought that bone-marrow derived stem cells are preferred for bone, cartilage, ligament, and tendon repair, others believe that the less challenging collection techniques and the multi-cellular microenvironment already present in adipose-derived stem cell fractions make the latter the preferred source for autologous transplantation.

Currently available treatments for horses and dogs suffering from orthopaedic conditions

Autologous or allogeneic stem cells are currently used as an adjunctive therapy in the surgical repair of some types of fractures in dogs and horses. Autologous stem cell-based treatments for ligament injury, tendon injury, osteoarthritis, osteochondrosis, and sub-chondral bone cysts have been commercially available to practicing veterinarians to treat horses since 2003 in the United States and since 2006 in the United

Kingdom. Autologous stem-cell based treatments for tendon injury, ligament injury, and osteoarthritis in dogs have been available to veterinarians in the United States since 2005. Over 3000 privately-owned horses and dogs have been treated with autologous adipose-derived stem cells. The efficacy of these treatments has been shown in double-blind clinical trials for dogs with osteoarthritis of the hip and elbow and horses with tendon damage. The efficacy of using stem cells, whether adipose-derived or bone-marrow derived, for treating tendon and ligament injuries in horses has support in the veterinary literature.

Developments in Stem Cell Treatments in Veterinary Internal Medicine Currently, research is being conducted to develop stem cell treatments for: 1) horses suffering from COPD, neurologic disease, and laminitis; and 2) dogs and cats suffering from heart disease, liver disease, kidney disease, neurologic disease, and immune-mediated disorders.

Embryonic Stem Cell Controversy

There is widespread controversy over the use of human embryonic stem cells. This controversy primarily targets the techniques used to derive new embryonic stem cell lines, which often requires the destruction of the blastocyst. Opposition to the use of human embryonic stem cells in research is often based on philosophical, moral or religious objections.

Stem Cell Treatments Around the World

China

Stem cell research and treatment is currently being practiced at a clinical level in the People's Republic of China. The Ministry of Health of the People's Republic of China has permitted the use of stem cell therapy for conditions beyond those approved of in Western countries such as the United States, United Kingdom, and Australia. The Western World has scrutinized China for its failed attempts to meet international documentation standards of these trials and procedures, despite the overwhelmingly positive anecdotal results.

Stem cell therapies provided in China utilize a variety of cell types including umbilical cord stem cells and olfactory ensheathing cells. The stem cells are then expanded in centralized blood banks before being used in stem cell treatments.

State-funded companies based in the Shenzhen Hi-Tech Industrial Zone treat the symptoms of numerous disorders with adult stem cell therapy. Hospitals throughout eastern China provide numerous therapies to patients in coordination with the stem cell providers.

These companies' therapies are currently focused on the treatment of neurodegenerative and cardiovascular disorders. The most radical successes of Chinese adult stem cell therapy have been in treating the brain. These therapies administer stem cells directly to the brain to promote greater motor and brain function in patients with Cerebral Palsy, Alzheimer's, and brain injuries.

However, retrospective studies have shown that Chinese use of fetal-derived brain tissue in spinal cord injured human subjects were not as promising as once thought: the phenotype and the fate of the transplanted cells, described as olfactory ensheathing cells, were unknown. As well, perioperative morbidity and lack of functional benefit were identified as the most serious clinical shortcomings. Furthermore, the extent of regulatory policy in the use of stem cell therapies in China is unclear. In the absence of a valid clinical trials protocol, and more regulatory oversight, Western regulatory agencies advise patients and physicians to be cautious when selecting Chinese stem cell therapeutic centers.

Mexico

Stem cell treatment is currently being practiced at a clinical level in Mexico. An International Health Department Permit (COFEPRIS) is required. This permit allows the use of stem cell types beyond those approved of in Western countries such as the United States or Europe. Stem cell therapies provided in Mexico utilize patient Adipose, Bone Marrow, or Donor Placenta sources.

South Korea

In 2005, South Korean scientists claimed to have generated stem cells that were tailored to match the recipient. Each of the 11 new stem cell lines was developed using somatic cell nuclear transfer (SCNT) technology. The resultant cells were thought to match the genetic material of the recipient, thus suggesting minimal to no cell rejection. This study, however, was eventually discredited as the primary researcher,Dr. Woo Suk Hwang, admitted to using cells obtained from his research staff. In Dec 2005, claims were put forward that his research had been manipulated to wrongfully indicate positive results. Later that month, these claims were confirmed by an academic panel.

Ukraine

Today, Ukraine is permitted to perform clinical trials of stem cell treatments (Order of the MH of Ukraine [1] 630 "About carrying out clinical trials of stem cells", 2008) for the treatment of these pathologies: pancreatic necrosis, cirrhosis, hepatitis, burn disease, diabetes, multiple sclerosis, critical lower limb ischemia. The first medical institution granted the right to conduct clinical trials became the "Institute of Cell Therapy"(Kiev).

- Cardiovascular Cell Therapy Research Network (CCTRN)
- Fetal tissue implant
- Human Stem Cells Institute

Pharmacological

As of August 2008, several trials of various compounds are in development or ongoing, a few at the point of testing on larger numbers of people, known as phase III of clinical trials. Substances that have shown promise in initial experiments include dopamine receptor blockers, select dopamine antagonists, such as creatine and CoQ10, the antibiotic Minocycline, antioxidant-containing foods and nutrients, and antidepressants, including selective serotonin reuptake inhibitors such as sertraline, fluoxetine, and paroxetine. In 2010, minocycline was found to be ineffective for humans in a multi-center trial. A 2009 study suggested strong association between specific single nucleotide polymorphisms alleles and CAG expansion also provides an opportunity of personalized therapeutics in HD where the clinical development of only a small number of allele-specific targets may be sufficient to treat up to 88% of the HD patient population.

Other

Recently it was shown that increasing BDNF expression may represent a useful strategy for Huntington's disease treatment.

- Gauchers disease is a genetic disease affecting metabolism. It is more treatable then most other genetic diseases, and can be treated with enzyme replacement therapy, medication miglustat, and bone marrow transplantion.

Classification

Gaucher's disease has three common clinical subtypes.

- Type I (or non-neuropathic type) is the most common form of the disease, occurring in approximately 1 in 50,000 live births. It occurs most often among persons of Ashkenazi Jewish heritage. Symptoms may begin early in life or in adulthood and include enlarged liver and grossly enlarged spleen (together hepatosplenomegaly); the spleen can rupture and cause additional complications. Skeletal weakness and bone disease may be extensive. Spleen enlargement and bone marrow replacement cause anemia, thrombocytopenia and leukopenia. The brain is not affected pathologically, but there may be lung and, rarely, kidney impairment. Patients in this group usually bruise easily (due to low levels of platelets) and experience fatigue due to low numbers of red blood cells. Depending on disease onset and severity, type 1 patients may live well into adulthood. Many patients have a mild form of the disease or may not show any symptoms.
- Type II (or acute infantile neuropathic Gaucher's disease) typically begins within 6 months of birth and has an incidence rate of approximately 1 in 100,000 live births. Symptoms include an enlarged liver and spleen, extensive and progressive brain damage, eye movement disorders, spasticity, seizures, limb rigidity, and a poor ability to suck and swallow. Affected children usually die by age 2.
- Type III (the chronic neuropathic form) can begin at any time in childhood or even in adulthood, and occurs

in approximately 1 in 100,000 live births. It is characterized by slowly progressive but milder neurologic symptoms compared to the acute or type 2 version. Major symptoms include an enlarged spleen and/or liver, seizures, poor coordination, skeletal irregularities, eye movement disorders, blood disorders including anemia and respiratory problems. Patients often live into their early teen years and adulthood.

These subtypes have come under some criticism for not taking account of the full spectrum of observable symptoms (the phenotypes.) There are also compound heterozygous variations which considerably increase the complexity of predicting disease course.

Signs and Symptoms

- Painless hepatomegaly and splenomegaly: the size of the spleen can be 1500-3000 ml, as opposed to the normal size of 50-200 ml. Splenomegaly may decrease the affected individual's capacity for eating by exerting pressure on the stomach. While painless, enlargement of spleen increases the risk of splenic rupture.
- Hypersplenism and pancytopenia: the rapid and premature destruction of blood cells, leading to anemia, neutropenia, leukopenia, and thrombocytopenia (with an increased risk of infection and bleeding).
- Cirrhosis of the liver is rare
- Severe pain associated with joints and bones, frequently presenting in hips and knees.
- Neurological symptoms occur only in some types of Gaucher's:
  - Type II: serious convulsions, hypertonia, mental retardation, apnea.
  - Type III: muscle twitches known as myoclonus, convulsions, dementia, ocular muscle apraxia.
- Osteoporosis: 75% develop visible bony abnormalities due to the accumulated glucosylceramide. A deformity of the distal femur in the shape of an Erlenmeyer flask is commonly described (aseptic necrosis of the femur joint).
- Yellowish-brown skin pigmentation

Pathophysiology

The disease is caused by a defect in housekeeping gene lysosomal gluco-cerebrosidase (also known as beta-glucosidase, EC 3.2.1.45, PDB 1OGS) on the first chromosome (1q21). The enzyme is a 55.6 KD, 497 amino acids long protein that catalyses the breakdown of glucosylceramidase, a cell membrane constituent of red and white blood cells. The macrophages that clear these cells are unable to eliminate the waste product, which accumulates in fibrils, and turn into Gaucher cells, which appear on light microscopy to resemble crumpled-up paper. In the brain (type II and III), glucosylceramidase accumulates due to the turnover of complex lipids during brain development and the formation of the myelin sheath of nerves. Different mutations in the beta-glucosidase determine the remaining activity of the enzyme, and, to a large extent, the phenotype. Heterozygotes for particular acid beta-glucosidase mutations carry about a fivefold risk of developing Parkinson's disease, making this the most common known genetic risk-factor for Parkinson's. A study of 1525 Gaucher patients in the United States suggested that while cancer risk is not elevated, particular malignancies (non-Hodgkin lymphoma, melanoma and pancreatic cancer) occurred at a 2-3 times higher rate.

Genetics

The three types of Gaucher's disease are inherited in an autosomal recessive fashion. Both parents must be carriers in order for a child to be affected. If both parents are carriers, there is a one in four, or 25%, chance with each pregnancy for an affected child. Genetic counseling and genetic testing is recommended for families who may be carriers of mutations. Each type has been linked to particular mutations. In all, there are about 80 known mutations, grouped into three main types:

- Type I (N370S homozygote), the most common, also called the "non-neuropathic" type occurs mainly in Ashkenazi Jews, at 100 times the occurrence in the general populace. The median age at diagnosis is 28 years of age, and life expectancy is mildly decreased. There are no neurological symptoms.
- Type II (1 or 2 alleles L444P) is characterized by neurological problems in small children. The enzyme is

hardly released into the lysosomes. Prognosis is dismal: most die before reaching the third birthday.

- Type III (also 1-2 copies of L444P, possibly delayed by protective polymorphisms) occurs in Swedish patients from the Norrbotten region. This group develops the disease somewhat later, but most die before their 30th birthday.

Diaz et al. suggest that the Gaucher-causing mutations entered the Ashkenazi Jewish gene pool in the early Middle Ages (48-55 generations ago).

Diagnosis

Micrograph showing crinkled paper macrophages in the marrow space in a case of Gaucher disease. H&E stain. A definitive diagnosis is made with genetic testing. As there are numerous different mutations, sequencing of the beta-glucosidase gene is sometimes necessary to confirm the diagnosis. Prenatal diagnosis is available, and is useful when there is a known genetic risk factor. A diagnosis can also be implied by biochemical abnormalities such as high alkaline phosphatase, angiotensin-converting enzyme (ACE) and immunoglobulin levels, or by cell analysis showing "crinkled paper" cytoplasm and glycolipid-laden macrophages. Some lysosomal enzymes are elevated, including tartrate-resistant acid phosphatase, hexosaminidase, and a human chitinase, chitotriosidase. This latter enzyme has proved to be very useful for monitoring Gaucher's disease activity in response to treatment, and may reflect the severity of the disease

Treatment

For type 1 and most type 3 patients, enzyme replacement treatment with intravenous recombinant glucocerebrosidase (imiglucerase) can dramatically decrease liver and spleen size, reduce skeletal abnormalities, and reverse other manifestations. This treatment costs approximately $200,000 annually for a single patient and should be continued for life. The rarity of the disease means that dose-finding studies have been difficult to conduct, so there remains controversy over the optimal dose and dosing frequency. Due to the low incidence, this has become an orphan drug in many countries, meaning that a government recognizes and accommodates the financial constraints that limit research into drugs that address a small population. Velaglucerase alfa was approved by the Food and Drug Administration (FDA) as an alternative treatment on February 26, 2010.

Successful bone marrow transplantation cures the non-neurological manifestations of the disease, because it introduces a monocyte population with active beta-glucosidase. However, this procedure carries significant risk and is rarely performed in Gaucher patients. Surgery to remove the spleen (splenectomy) may be required on rare occasions if the patient is anemic or when the enlarged organ affects the patient's comfort. Blood transfusion may benefit some anemic patients. Other patients may require joint replacement surgery to improve mobility and quality of life. Other treatment options include antibiotics for infections, antiepileptics for seizures, bisphosphonates for bone lesions, and liver transplants. Substrate reduction therapy may prove to be effective in stopping Type 2, as it can cross through the blood barrier into the brain. There is currently no effective treatment for the severe brain damage that may occur in patients with types 2 and 3 Gaucher disease. Gene therapy may be a future step. Gaucher's disease has recently become a target for more than one effort at pharmacological chaperoning, which involves the use of orally administered drugs that operate at a molecular level. Miglustat is one of these oral drugs. It was approved for the treatment of this disease in 2003. As of June 2009, another oral drug, isofagomine tartrate, is under development.

Epidemiology

The National Gaucher Foundation states that around 1 in 100 people in the general U.S. population is a carrier for type 1 Gaucher's disease, giving a prevalence of 1 in 40,000: among Ashkenazi Jews the rate of carriers is considerably higher, at roughly 1 in 15. Type 2 Gaucher's disease shows no particular preference for any ethnic group. Type 3 Gaucher's disease is especially common in the population of the Northern Swedish region of Norrbotten where the incidence of the disease is 1 in 50,000.

History

The disease was first recognised by the French doctor Philippe Gaucher, who originally described it in 1882 and lent his name to the condition. The biochemical basis for the disease would be elucidated in 1965. The first effective treatment for the disease, the drug Ceredase, was approved by the FDA in April 1991. An improved drug, Cerezyme, was approved by the FDA in May 1994 and has replaced the use of Ceredase.

Genetic Epidemiology

Genetic epidemiology is the study of the role of genetic factors in determining health and disease in families and in populations, and the interplay of such genetic factors with environmental factors. In slightly more formal language, genetic epidemiology was defined by Morton as "a science which deals with the etiology, distribution, and control of disease in groups of relatives and with inherited causes of disease in populations". It is closely allied to both molecular epidemiology and statistical genetics, but these overlapping fields each have distinct emphases, societies and journals. Traditionally, the study of the role of genetics in disease progresses through the following study designs, each answering a slightly different question:

- Familial aggregation studies: Is there a genetic component to the disease, and what are the relative contributions of genes and environment?
- Segregation studies: What is the pattern of inheritance of the disease (e.g. dominant or recessive)?
- Linkage studies: On which part of which chromosome is the disease gene located?
- Association studies: Which allele of which gene is associated with the disease?

This traditional approach has proved highly successful in identifying monogenic disorders and locating the genes responsible. More recently, the scope of genetic epidemiology has expanded to include common diseases for which many genes each make a smaller contribution (polygenic, multifactorial or multigenic disorders).

This has developed rapidly in the first decade of the 21st century following completion of the Human Genome Project, as advances in genotyping technology and associated reductions in cost has made it feasible to conduct large-scale genome-wide association studies that genotype many thousands of single nucleotide polymorphisms in thousands of individuals. These have led to the discovery of many genetic polymorphisms that influence the risk of developing many common diseases.

Genotoxicity Bioindicator

Cluster analysis in epidemiology is the method for identifying wide spread population damage (as mercury contamination in Minamata Bay) but a genotoxicity bioindicator is a preventive way of seeing congenital malformations epidemic in human populations. A biological monitor, or biomonitor, is defined as an organism that provides quantitative information on the quality of the environment around it. Therefore, a good biomonitor will indicate the presence of the pollutant and also attempt to provide additional information about the amount and intensity of the exposure. Endocrine Disruption (feminization); Reproductive Failure causes include early embryonic death, embryo toxicity and abnormal parental behaviour during incubation; Eggshell Thinning including Double-crested Cormorants, Ospreys, Bald Eagles, Black-crowned Night-Herons and Herring Gulls, maybe are caused by pollution or another genotoxic factor.

Malformed frogs are caused by parasites, genetic disease or pollution? Crossed bills, jaw defects, extra limbs, and malformed feet, joints and eyes were found in Herring Gulls and at least eight other species of fish-eating waterbirds from great lakes, malformed frogs were found in Minnesota.

Inborn Errors of Metabolism

Inborn errors of metabolism comprise a large class of genetic diseases involving disorders of metabolism. The majority are due to defects of single genes that code for enzymes that facilitate conversion of various substances (substrates) into others (products). In most of the disorders, problems arise due to accumulation of substances which are toxic or interfere with normal function, or to the effects of reduced ability to synthesize essential compounds. Inborn errors of metabolism are now often referred to as congenital metabolic diseases or inherited metabolic diseases. The term inborn error of metabolism was coined by a British physician, Archibald Garrod (1857–1936), in the early 20th century (1908). He is known for work that prefigured the "one gene-one enzyme" hypothesis, based on his studies on the nature and inheritance of alkaptonuria. His seminal text, Inborn Errors of Metabolism was published in 1923.

Major Categories of Inherited Metabolic Diseases

Traditionally the inherited metabolic diseases were categorized as disorders of carbohydrate metabolism, amino acid metabolism, organic acid metabolism, or lysosomal storage diseases. In recent decades, hundreds of new inherited disorders of metabolism have been discovered and the categories have proliferated. Following are some of the major classes of congenital metabolic diseases, with prominent examples of each class. Many others do not fall into these categories. ICD-10 codes are provided where available.

- Disorders of carbohydrate metabolism
  - E.g., glycogen storage disease
- Disorders of amino acid metabolism
  - E.g., phenylketonuria , maple syrup urine disease, glutaric acidemia type 1
- Disorders of organic acid metabolism (organic acidurias)
  - E.g., alcaptonuria
- Disorders of fatty acid oxidation and mitochondrial metabolism
  - E.g., Medium-chain acyl-coenzyme A dehydrogenase deficiency (often shortened to MCADD.)
- Disorders of porphyrin metabolism
  - E.g., acute intermittent porphyria
- Disorders of purine or pyrimidine metabolism
  - E.g., Lesch-Nyhan syndrome
- Disorders of steroid metabolism
  - E.g., congenital adrenal hyperplasia
- Disorders of mitochondrial function
  - E.g., Kearns-Sayre syndrome
- Disorders of peroxisomal function
  - E.g., Zellweger syndrome
- Lysosomal storage disorders
  - E.g., Gaucher's disease
  - E.g., Niemann Pick disease

Manifestations and Presentations

Because of the enormous number of these diseases and wide range of systems affected, nearly every "presenting complaint" to a doctor may have a congenital metabolic disease as a possible cause, especially in childhood. The following are examples of potential manifestations affecting each of the major organ systems: many manifestations may develop

- Growth failure, failure to thrive, weight loss
- Ambiguous genitalia, delayed puberty, precocious puberty
- Developmental delay, seizures, dementia, encephalopathy, stroke
- Deafness, blindness, pain agnosia
- Skin rash, abnormal pigmentation, lack of pigmentation, excessive hair growth, lumps and bumps
- Dental abnormalities
- Immunodeficiency, thrombocytopenia, anemia, enlarged spleen,
  enlarged lymph nodes
- Many forms of cancer
- Recurrent vomiting, diarrhea, abdominal pain
- Excessive urination, renal failure, dehydration, edema
- Hypotension, heart failure, enlarged heart, hypertension, myocardial infarction
- Hepatomegaly, jaundice, liver failure
- Unusual facial features, congenital malformations
- Excessive breathing (hyperventilation), respiratory failure
- Abnormal behavior, depression, psychosis
- Joint pain, muscle weakness, cramps
- Hypothyroidism, adrenal insufficiency, hypogonadism, diabetes mellitus

Diagnostic Techniques

Dozens of congenital metabolic diseases are now detectable by newborn screening tests, especially the expanded testing using mass spectrometry. This is an increasingly common way for the diagnosis to be made and sometimes results in earlier treatment and a better outcome. There is a revolutionary GC/MS based technology with an integrated analytics system, which has now made it possible to test a newborn for over 100 genetic metabolic disorders. Because of the multiplicity of conditions, many different diagnostic tests are used for screening. An abnormal result is often followed by a subsequent "definitive test" to confirm the suspected diagnosis.

Common screening tests used in the last sixty years:

- Ferric chloride test (turned colors in reaction to various abnormal metabolites in urine)
- Ninhydrin paper chromatography (detected abnormal amino acid patterns)
- Guthrie bacterial inhibition assay (detected a few amino acids in excessive amounts in blood) The dried blood spot can be used for multianalyte testing using Tandem Mass Spectrometry (MS/MS). This given an indication for a disorder. The same has to be further confirmed by enzyme assays, GC/MS or DNA Testing.
- Quantitative measurement of amino acids in plasma and urine
- Urine organic acid analysis by Gas chromatography-mass spectrometry
- Plasma acylcarnitines analysis by mass spectrometry
- Urine purines and pyrimidines analysis by Gas chromatography-mass spectrometry

Specific diagnostic tests (or focused screening for a small set of disorders):

- Tissue biopsy or necropsy: liver, muscle, brain, bone marrow
- Skin biopsy and fibroblast cultivation for specific enzyme testing
- Specific DNA testing

Treatment

In the middle of the 20th century the principal treatment for some of the amino acid disorders was restriction of dietary protein and all other care was simply management of complications. In the past twenty years, enzyme replacement, gene transfer, and organ transplantation have become available and beneficial for many previously untreatable disorders. Some of the more common or promising therapies are listed:

- Dietary restriction
  - E.g., reduction of dietary protein remains a mainstay of treatment for phenylketonuria and other amino acid disorders
- Dietary supplementation or replacement
  - E.g., oral ingestion of cornstarch several times a day helps prevent people with glycogen storage diseases from becoming seriously hypoglycemic.
- Vitamins
  - E.g., thiamine supplementation benefits several types of disorders that cause lactic acidosis.
- Intermediary metabolites, compounds, or drugs that facilitate or retard specific metabolic pathways
- Dialysis
- Enzyme replacement E.g. Acid-alpha glucosidase for Pompe disease
- Gene transfer
- Bone marrow or organ transplantation
- Treatment of symptoms and complications
- Prenatal diagnosis and avoidance of pregnancy or abortion of an affected fetus

Resources

For clinicians and scientists in the field of inborn errors of metabolism, good resources include books by Scriver, Fernandes, Clarke, Blau (diagnosis), Blau (treatment), Lyon, Nyhan, Hoffmann, and Zschocke. Other resources include genetests, orphanet, OMIM, Metab-L, societies such as the SSIEM, the SIMD and links therein. For medical students and clinicians looking for overviews of the field, such reviews can be found on pubmed and in good pediatric textbooks (e.g. articles by Saudubray, Ellaway, Raghuveer or Burton, and textbooks by Hay or Behrman). For patients, their families or other individuals seeking good information and support groups, the National Institutes of Health offers the office of rare diseases, genetics home reference, medlineplus and health information.

The National Human Genome Research Institute hosts an information center, a section for patients and the public and additional educational resources. Support groups can be found at NORD, Genetic Alliance and Orphanet. The genetic education center at the KUMC has many more useful links.

Medical Genetics

Medical genetics is the specialty of medicine that involves the diagnosis and management of hereditary disorders. Medical genetics differs from Human genetics in that human genetics is a field of scientific research that may or may not apply to medicine, but medical genetics refers to the application of genetics to medical care. For example, research on the causes and inheritance of genetic disorders would be considered within both human genetics and medical genetics, while the diagnosis, management, and counseling of individuals with genetic disorders would be considered part of medical genetics.

In contrast, the study of typically non-medical phenotypes such as the genetics of eye color would be considered part of human genetics, but not necessarily relevant to medical genetics (except in situations such as albinism). Genetic medicine is a newer term for medical genetics and incorporates areas such as gene therapy, personalized medicine, and the rapidly emerging new medical specialty, predictive medicine.

Scope

Medical genetics encompasses many different areas, including clinical practice of physicians, genetic counselors, and nutritionists, clinical diagnostic laboratory activities, and research into the causes and inheritance of genetic disorders. Examples of conditions that fall within the scope of medical genetics include birth defects and dysmorphology, mental retardation, autism, and mitochondrial disorders, skeletal dysplasia, connective tissue disorders, cancer genetics, teratogens, and prenatal diagnosis. Medical genetics is increasingly becoming relevant to many common diseases. Overlaps with other medical specialties are beginning to emerge, as recent advances in genetics are revealing etiologies for neurologic, endocrine, cardiovascular, pulmonary, ophthalmologic, renal, psychiatric, and dermatologic conditions.

Subspecialties

In some ways, many of the individual fields within medical genetics are hybrids between clinical care and research. This is due in part to recent advances in science and technology (for example, the Human genome project) that have enabled an unprecedented understanding of genetic disorders.

Clinical Genetics

Clinical genetics is the practice of clinical medicine with particular attention to hereditary disorders. Referrals are made to genetics clinics for a variety of reasons, including birth defects, developmental delay, autism, epilepsy, short stature, and many others. Examples of genetic syndromes that are commonly seen in the genetics clinic include chromosomal rearrangements, Down syndrome, DiGeorge syndrome (22q11.2 Deletion Syndrome), Fragile X syndrome, Marfan syndrome, Neurofibromatosis, Turner syndrome, and Williams syndrome.

Metabolic/Biochemical Genetics

Metabolic (or biochemical) genetics involves the diagnosis and management of inborn errors of metabolism in which patients have enzymatic deficiencies that perturb biochemical pathways involved in metabolism of carbohydrates, amino acids, and lipids. Examples of metabolic disorders include galactosemia, glycogen storage disease, lysosomal storage disorders, metabolic acidosis, peroxisomal disorders, phenylketonuria, and urea cycle disorders.

Cytogenetics

Cytogenetics is the study of chromosomes and chromosome abnormalities. While cytogenetics historically relied on microscopy to analyze chromosomes, new molecular technologies such as array comparative genomic hybridization are now becoming widely used. Examples of chromosome abnormalities include aneuploidy, chromosomal rearrangements, and genomic deletion/duplication disorders.

Molecular Genetics

Molecular genetics involves the discovery of and laboratory testing for DNA mutations that underlie many single gene disorders. Examples of single gene disorders include achondroplasia, cystic fibrosis, Duchenne muscular dystrophy, hereditary breast cancer (BRCA1/2), Huntington disease, Marfan syndrome, Noonan syndrome, and Rett syndrome. Molecular tests are also used in the diagnosis of syndromes involving epigenetic abnormalities, such as Angelman syndrome, Beckwith-Wiedemann syndrome, Prader-willi syndrome, and uniparental disomy.

Mitochondrial genetics

Mitochondrial genetics concerns the diagnosis and management of mitochondrial disorders, which have a molecular basis but often result in biochemical abnormalities due to deficient energy production. There exists some overlap between medical genetic diagnostic laboratories and molecular pathology.

Genetic Counseling

Genetic counseling is the process of providing information about genetic conditions, diagnostic testing, and risks in other family members, within the framework of nondirective counseling. Genetic counselors are non-physician members of the medical genetics team who specialize in family risk assessment and counseling of patients regarding genetic disorders. The precise role of the genetic counselor varies somewhat depending on the disorder. Although genetics has its roots back in the 19th century with the work of the Bohemian monk Gregor Mendel and other pioneering scientists, human genetics emerged later. It started to develop, albeit slowly, during the first half of the 20th century. Mendelian (single-gene) inheritance was studied in a number of important disorders such as albinism, brachydactyly (short fingers and toes), and hemophilia. Mathematical approaches were also devised and applied to human genetics.

Population genetics was created. Medical genetics was a late developer, emerging largely after the close of World War II (1945) when the eugenics movement had fallen into disrepute. The Nazi misuse of eugenics sounded its death knell.

Shorn of eugenics, a scientific approach could be used and was applied to human and medical genetics. Medical genetics saw an increasingly rapid rise in the second half of the 20th century and continues in the 21st century.

Notable Practitioners

- Victor McKusick (1921–2008), "The Father of Genetic Medicine" and principal author of Mendelian Inheritance in Man.

Current Practice

The clinical setting in which patients are evaluated determines the scope of practice, diagnostic, and therapeutic interventions. For the purposes of general discussion, the typical encounters between patients and genetic practitioners may involve:

- Referral to an out-patient genetics clinic (pediatric, adult, or combined) or an in-hospital consultation, most often for diagnostic evaluation.
- Specialty genetics clinics focusing on management of inborn errors of metabolism, skeletal dysplasia, or lysosomal storage diseases.
- Referral for counseling in a prenatal genetics clinic to discuss risks to the pregnancy (advanced maternal age, teratogen exposure, family history of a genetic disease), test results (abnormal maternal serum screen, abnormal ultrasound), and/or options for prenatal diagnosis (typically amniocentesis or chorionic villus sampling).
- Multidisciplinary specialty clinics that include a clinical geneticist or genetic counselor (cancer genetics, cardiovascular genetics, craniofacial or cleft lip/palate, hearing loss clinics, muscular dystrophy/neurodegenerative disorder clinics).

Diagnostic Evaluation

Each patient will undergo a diagnostic evaluation tailored to their own particular presenting signs and symptoms. The geneticist will establish a differential diagnosis and recommend appropriate testing.

Increasingly, clinicians use SimulConsult, paired with the National Library of Medicine Gene Review articles, to narrow the list of hypotheses (known as the differential diagnosis) and identify the tests that are relevant for a particular patient. These tests might evaluate for chromosomal disorders, inborn errors of metabolism, or single gene disorders.

Chromosome Studies

Chromosome studies are used in the general genetics clinic to determine a cause for developmental delay/mental retardation, birth defects, dysmorphic features, and/or autism. Chromosome analysis is also performed in the prenatal setting to determine whether a fetus is affected with aneuploidy or other chromosome rearrangements. Finally, chromosome abnormalities are often detected in cancer samples. A large number of different methods have been developed for chromosome analysis:

- Chromosome analysis using a karyotype involves special stains that generate light and dark bands, allowing identification of each chromosome under a microscope.
- Fluorescence in situ hybridization (FISH) involves fluorescent labeling of probes that bind to specific DNA sequences, used for identifying aneuploidy, genomic deletions or duplications, characterizing chromosomal translocations and determining the origin of ring chromosomes.
- Chromosome painting is a technique that uses fluorescent probes specific for each chromosome to differentially label each chromosome. This technique is more often used in cancer cytogenetics, where complex chromosome rearrangements can occur.
- Array comparative genomic hybridization is a new molecular technique that involves hybridization of an individual DNA sample to a glass slide or microarray chip containing molecular probes (ranging from large ~200kb bacterial artificial chromosomes to small oligonucleotides) that represent unique regions of the genome. This method is particularly sensitive for detection of genomic gains or losses across the genome but does not detect balanced translocations or distinguish the location of duplicated genetic material (for example, a tandem duplication versus an insertional duplication).

Basic Metabolic Studies

Biochemical studies are performed to screen for imbalances of metabolites in the bodily fluid, usually the blood (plasma/ serum) or urine, but also in cerebrospinal fluid (CSF). Specific tests of enzyme function (either in leukocytes, skin fibroblasts, liver, or muscle) are also employed under certain circumstances. In the US, the newborn screen incorporates biochemical tests to screen for treatable conditions such as galactosemia and phenylketonuria (PKU).

Patients suspected to have a metabolic condition might undergo the following tests:

- Quantitative amino acid analysis is typically performed using the ninhydrin reaction, followed by liquid chromatography to measure the amount of amino acid in the sample (either urine, plasma/serum, or CSF). Measurement of amino acids in plasma or serum is used in the evaluation of disorders of amino acid metabolism such as urea cycle disorders, maple syrup urine disease, and PKU. Measurement of amino acids in urine can be useful in the diagnosis of cystinuria or renal Fanconi syndrome as can be seen in cystinosis.
- Urine organic acid analysis can be either performed using quantitative or qualitative methods, but in either case the test is used to detect the excretion of abnormal organic acids. These compounds are normally produced during bodily metabolism of amino acids and odd-chain fatty acids, but accumulate in patients with certain metabolic conditions.
- The acylcarnitine combination profile detects compounds such as organic acids and fatty acids conjugated to carnitine. The test is used for detection of disorders involving fatty acid metabolism, including MCAD.
- Pyruvate and lactate are byproducts of normal metabolism, particularly during anaerobic metabolism. These compounds normally accumulate during exercise or ischemia, but are also elevated in patients with disorders of pyruvate metabolism or mitochondrial disorders.
- Ammonia is an end product of amino acid metabolism and is converted in the liver to urea through a series of enzymatic reactions termed the urea cycle. Elevated ammonia can therefore be detected in patients with urea cycle disorders, as well as other conditions involving liver failure.

- Enzyme testing is performed for a wide range of metabolic disorders to confirm a diagnosis suspected based on screening tests.

Molecular Studies

- DNA sequencing is used to directly analyze the genomic DNA sequence of a particular gene. In general, only the parts of the gene that code for the expressed protein (exons) and small amounts of the flanking untranslated regions and introns are analyzed. Therefore, although these tests are highly specific and sensitive, they do not routinely identify all of the mutations that could cause disease.
- DNA methylation analysis is used to diagnose certain genetic disorders that are caused by disruptions of epigenetic mechanisms such as genomic imprinting and uniparental disomy.
- Southern blotting is an early technique basic on detection of fragments of DNA separated by size through gel electrophoresis and detected using radiolabeled probes. This test was routinely used to detect deletions or duplications in conditions such as Duchenne muscular dystrophy but is being replaced by high-resolution array comparative genomic hybridization techniques. Southern blotting is still useful in the diagnosis of disorders caused by trinucleotide repeats.
- Short tandem repeats are unique markers that can be used to determine haplotypes and are used in identity testing for maternal cell contamination.

Treatments

Each cell of the body contains the hereditary information (DNA) wrapped up in structures called chromosomes. Since genetic syndromes are typically the result of alterations of the chromosomes or genes, there is no treatment currently available that can correct the genetic alterations in every cell of the body. Therefore, there is currently no "cure" for genetic disorders. However, for many genetic syndromes there is treatment available to manage the symptoms. In some cases, particularly inborn errors of metabolism, the mechanism of disease is well understood and offers the potential for dietary and medical management to prevent or reduce the long-term complications. In other cases, infusion therapy is used to replace the missing enzyme. Current research is actively seeking to use gene therapy or other new medications to treat specific genetic disorders.

Diet

Dietary restriction and supplementation are key measures taken in several well-known metabolic disorders, including galactosemia, phenylketonuria (PKU), maple syrup urine disease, organic acidurias and urea cycle disorders. Such restrictive diets can be difficult for the patient and family to maintain, and require close consultation with a nutritionist who has special experience in metabolic disorders. The composition of the diet will change depending on the caloric needs of the growing child and special attention is needed during a pregnancy if a woman is affected with one of these disorders.

Medication

Medical approaches include enhancement of residual enzyme activity (in cases where the enzyme is made but is not functioning properly), inhibition of other enzymes in the biochemical pathway to prevent buildup of a toxic compound, or diversion of a toxic compound to another form that can be excreted. Examples include the use of high doses of pyridoxine (vitamin B6) in some patients with homocystinuria to boost the activity of the residual cystathione synthase enzyme, administration of biotin to restore activity of several enzymes affected by deficiency of biotinidase, treatment with NTBC in Tyrosinemia to inhibit the production of succinylacetone which causes liver toxicity, and the use of sodium benzoate to decrease ammonia build-up in urea cycle disorders.

Enzyme Replacement Therapy

Certain lysosomal storage diseases are treated with infusions of a recombinant enzyme (produced in a laboratory), which can reduce the accumulation of the compounds in various tissues. Examples include Gaucher disease, Fabry disease, Mucopolysaccharidoses and Glycogen storage disease type II. Such treatments are limited by the ability of the enzyme to reach the affected areas (the blood brain barrier prevents enzyme from reaching the brain, for example), and can sometimes be associated with allergic reactions. The long-term clinical effectiveness of enzyme replacement therapies vary widely among different disorders.

Chapter 9

Genetic Recombination

Structure of the Holliday junction intermediate in genetic recombination. The four separate DNA strands are coloured red, blue, green and yellow.

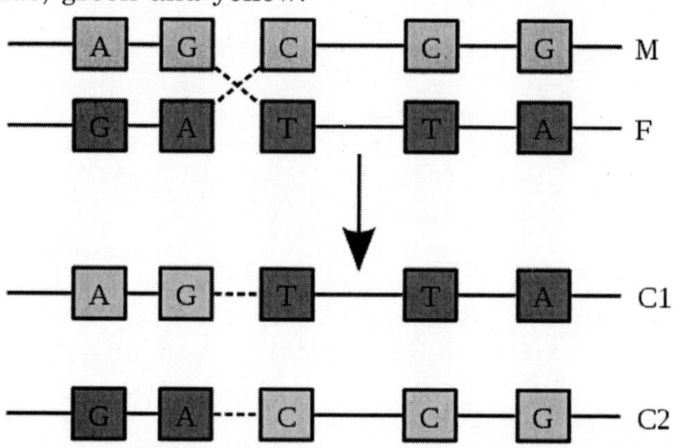

Figure: Recombination involves the breakage and rejoining of two chromosomes (M and F) to produce two re-arranged chromosomes (C1 and C2).

A DNA helix usually does not interact with other segments of DNA, and in human cells the different chromosomes even occupy separate areas in the nucleus called "chromosome territories". This physical separation of different chromosomes is important for the ability of DNA to function as a stable repository for information, as one of the few times chromosomes interact is during chromosomal crossover when they recombine. Chromosomal crossover is when two DNA helices break, swap a section and then rejoin. Recombination allows chromosomes to exchange genetic information and produces new combinations of genes, which increases the efficiency of natural selection and can be important in the rapid evolution of new proteins. Genetic recombination can also be involved in DNA repair, particularly in the cell's response to double-strand breaks.

The most common form of chromosomal crossover is homologous recombination, where the two chromosomes involved share very similar sequences. Non-homologous recombination can be damaging to cells, as it can produce chromosomal translocations and genetic abnormalities. The recombination reaction is catalyzed by enzymes known as recombinases, such as RAD51.

The first step in recombination is a double-stranded break either caused by an endonuclease or damage to the DNA. A series of steps catalyzed in part by the recombinase then leads to joining of the two helices by at least one Holliday junction, in which a segment of a single strand in each helix is annealed to the complementary strand in the other helix. The Holliday junction is a tetrahedral junction structure that can be moved along the pair of chromosomes, swapping one strand for another. The recombination reaction is then halted by cleavage of the junction and re-ligation of the released DNA.

Evolution

DNA contains the genetic information that allows all modern living things to function, grow and reproduce. However, it is unclear how long in the 4-billion-year history of life DNA has performed this function, as it has been proposed that the earliest forms of life may have used RNA as their genetic material. RNA may have acted as the central part of early cell metabolism as it can both transmit genetic information and carry out catalysis as part of ribozymes. This ancient RNA world where nucleic acid would have been used for both catalysis and genetics may have influenced the evolution of the current genetic code based on four nucleotide bases. This would occur, since the number of different bases in such an organism is a trade-off between a small number of bases increasing replication accuracy and a large number of bases increasing the catalytic efficiency of ribozymes. However, there is no direct evidence of ancient genetic systems, as recovery of DNA from most fossils is impossible. This is because DNA will survive in the environment for less than one million years and slowly degrades into short fragments in solution. Claims for older DNA have been made, most notably a report of the isolation of a viable bacterium from a salt crystal 250 million years old, but these claims are controversial.

On August 8, 2011, a report, based on NASA studies with meteorites found on Earth, was published suggesting building blocks of DNA (adenine, guanine and related organic molecules) may have been formed extraterrestrially in outer space.

Uses in Technology

Methods have been developed to purify DNA from organisms, such as phenol-chloroform extraction, and to manipulate it in the laboratory, such as restriction digests and the polymerase chain reaction. Modern biology and biochemistry make intensive use of these techniques in recombinant DNA technology. Recombinant DNA is a man-made DNA sequence that has been assembled from other DNA sequences. They can be transformed into organisms in the form of plasmids or in the appropriate format, by using a viral vector. The genetically modified organisms produced can be used to produce products such as recombinant proteins, used in medical research, or be grown in agriculture.

Forensics

Forensic scientists can use DNA in blood, semen, skin, saliva or hair found at a crime scene to identify a matching DNA of an individual, such as a perpetrator. This process is formally termed DNA profiling, but may also be called "genetic fingerprinting". In DNA profiling, the lengths of variable sections of repetitive DNA, such as short tandem repeats and minisatellites, are compared between people. This method is usually an extremely reliable technique for identifying a matching DNA. However, identification can be complicated if the scene is contaminated with DNA from several people. DNA profiling was developed in 1984 by British geneticist Sir Alec Jeffreys, and first used in forensic science to convict Colin Pitchfork in the 1988 Enderby murders case.

People convicted of certain types of crimes may be required to provide a sample of DNA for a database. This has helped investigators solve old cases where only a DNA sample was obtained from the scene. DNA profiling can also be used to identify victims of mass casualty incidents. On the other hand, many convicted people have been released from prison on the basis of DNA techniques, which were not available when a crime had originally been committed.

Bioinformatics

Bioinformatics involves the manipulation, searching, and data mining of biological data, and this includes DNA sequence data. The development of techniques to store and search DNA sequences have led to widely applied advances in computer science, especially string searching algorithms, machine learning and database theory. String searching or matching algorithms, which find an occurrence of a sequence of letters inside a larger sequence of letters, were developed to search for specific sequences of nucleotides. The DNA sequenced may be aligned with other DNA sequences to identify homologous sequences and locate the specific mutations that make them distinct. These techniques, especially multiple sequence alignment, are used in studying phylogenetic relationships and protein function.

Data sets representing entire genomes' worth of DNA sequences, such as those produced by the Human Genome Project, are difficult to use without the annotations that identify the locations of genes and regulatory elements on each chromosome. Regions of DNA sequence that have the characteristic patterns associated with protein- or RNA-coding genes can be identified by gene finding algorithms, which allow researchers to predict the presence of particular gene products and their possible functions in an organism even before they have been isolated experimentally.

Entire genomes may also be compared which can shed light on the evolutionary history of particular organism and permit the examination of complex evolutionary events.

DNA Nanotechnology

The DNA structure at left (schematic shown) will self-assemble into the structure visualized by atomic force microscopy at right. DNA nanotechnology is the field that seeks to design nanoscale structures using the molecular recognition properties of DNA molecules.

DNA nanotechnology uses the unique molecular recognition properties of DNA and other nucleic acids to create self-assembling branched DNA complexes with useful properties. DNA is thus used as a structural material rather than as a carrier of biological information. This has led to the creation of two-dimensional periodic lattices (both tile-based as well as using the "DNA origami" method) as well as three-dimensional structures in the shapes of polyhedra.

Nanomechanical devices and algorithmic self-assembly have also been demonstrated, and these DNA structures have been used to template the arrangement of other molecules such as gold nanoparticles and streptavidin proteins.

DNA Nanotechnology

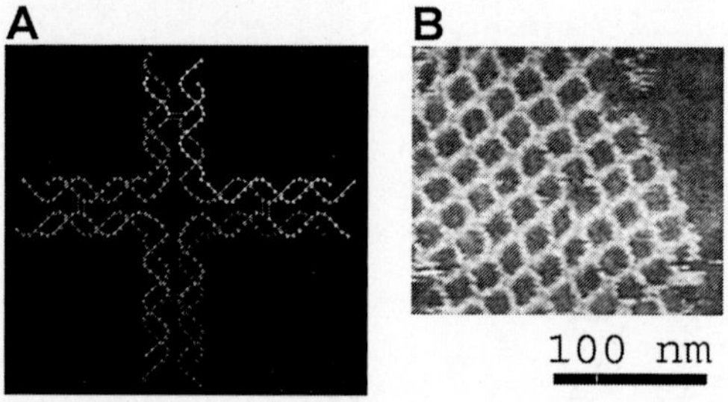

DNA nanotechnology is a branch of nanotechnology which uses the molecular recognition properties of DNA and other nucleic acids to create designed, artificial structures out of DNA for technological purposes. In this field, DNA is used as a structural material rather than as a carrier of genetic information, making it an example of bionanotechnology. DNA nanotechnology has applications in molecular self-assembly and in DNA computing. Although DNA is usually considered in the context of molecular biology as the carrier of genetic information in living cells, DNA nanotechnology considers DNA solely as a chemical and as a material, and is usually pursued outside of any biological context.

DNA nanotechnology makes use of the fact that, due to the specificity of Watson-Crick base pairing, only portions of the strands which are complementary to each other will bind to each other to form duplex DNA. DNA nanotechnology attempts to rationally design sets of DNA strands so that desired portions of each strand will assemble in the correct positions for some desired target structure, a process called nucleic acid design.

Although the field is usually called DNA nanotechnology, its principles apply equally well to other nucleic acids such as RNA and PNA, and structures incorporating these have been made. For this reason the field is occasionally referred to as nucleic acid nanotechnology.

Fundamental Concepts

These four strands associate into a DNA four-arm junction because this structure maximizes the number of correct base pairs, with A's matched to T's and C's matched to G's.

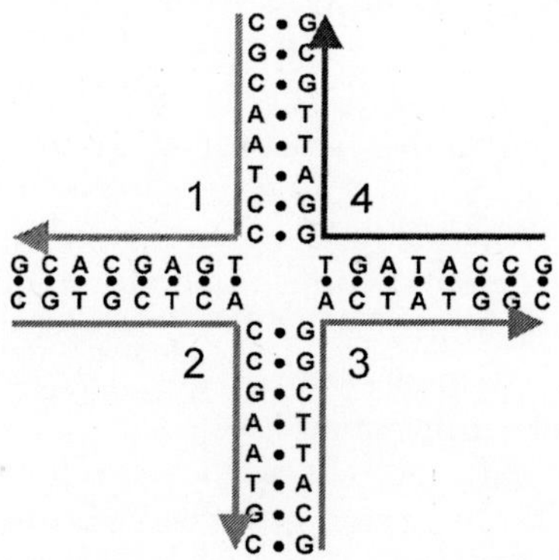

See this image for a more realistic model of the four-arm junction showing its tertiary structure.

A double-crossover (DX) molecule. This molecule consists of five DNA single strands which form two double-helical domains, on the left and the right in this image. There are two crossover points where the strands cross from one domain into the other.

DNA nanotechnology creates complex structures out of nucleic acids by making use of the specificity of base pairing in nucleic acid molecules. The structure of a nucleic acid molecule consists of a sequence of nucleotides, distinguished by which nucleobase they contain. In DNA, the four bases used are adenine (A), cytosine (C), guanine (G), and thymine (T). Nucleic acids have the property that two molecules will bind to each other to form a double helix only if the two sequences are complementary, meaning that they form matching sequences of base pairs, with A's only binding to T's, and C's only to G's.

Because the formation of correctly matched base pairs is energetically favorable, nucleic acid strands are expected in most cases to bind to each other in the conformation that maximizes the number of correctly paired bases. This property, that the sequence determines the pattern of binding and the overall structure, is used by the field of DNA nanotechnology in that sequences are artificially designed so that a desired structure is favored to form. Nearly all structures in DNA nanotechnology make use of branched DNA structures containing junctions, as opposed to most biological DNA which exists in a linear double helix form. One of the simplest branched structures, and the first made, is a four-arm junction which can be made using four individual DNA strands which are complementary to each other in the correct pattern. Unlike in natural Holliday junctions, in the artificial immobile four-arm junction shown below, the base sequence of each arm is different, meaning that the junction point is fixed in a certain position.

Junctions can be used in more complex molecules. One of the more widely used of these is the "double-crossover" or DX motif. A DX molecule can be thought of as two DNA duplexes positioned parallel to each other, with two crossover points where strands cross from one duplex into the other. Each junction point is itself topologically a four-arm junction. This molecule has the advantage that the junction points are now constrained to a single orientation as opposed to being flexible as in the four-arm junction. This makes the DX motif suitable as a structural building block for larger DNA complexes.

Design

DNA nanostructures must be rationally designed so that the individual nucleic acid strands will assemble into the desired structures. The design process of such nanostructures usually begins with the specification of a desired target structure and/or functionality. Then, the overall secondary structure of the target molecule is designed, meaning the arrangement of nucleic acid strands within the structure, and which portions of those strands should be bound to each other. The last step is the primary structure design, the specification of the actual base sequences of each nucleic acid strand.

Structural Design

The first step in designing a nucleic acid nanostructure is to decide how a given structure should be represented by a specific arrangement of nucleic acid strands. This design step thus determines the secondary structure, or the series of base pairs which hold the individual strands together in the desired shape. There are several approaches which have been demonstrated:

- Tile-based structures. This approach breaks the target structure into smaller units with strong binding between the strands contained in each unit, and relatively weaker interactions between the units. It is often used to make periodic lattices, but can also be used to implement algorithmic self-assembly, making them one platform for DNA computing.
- Folding structures. An alternative to the tile-based approach, folding approaches make the nanostructure out of a single long strand. This long strand can either have a designed sequence which folds due to its interactions with itself, or it can be folded into the desired shape by using shorter, "staple" strands. This latter method is called DNA origami, which allows the creation of two- and three-dimensional shapes at the nanoscale using DNA.
- Kinetic assembly. Recently, there has been interest in controlling the kinetics of DNA self-assembly, so that transient dynamics can also be programmed into the assembly. Such a method also has the advantage of proceeding isothermally and thus not requiring a thermal annealing step required by solely thermodynamic approaches.

Sequence Design

After any of the above approaches are used to design the secondary structure of a target molecule, an actual sequence of nucleotides must be devised which will form into the desired structure. Nucleic acid design is the process of assigning a specific nucleic acid base sequence to each strand so that they will associate into a desired conformation. Nucleic acid design is central to the field of DNA nanotechnology. Most methods seek to designing sequences so that the target structure is a thermodynamic minimum, and mis-assembled structures have higher energies and are thus disfavored. This is done either through heuristic methods such as sequence symmetry minimization and coding theory based approaches, or by explicitly using a full nearest-neighbor thermodynamic model.

Geometric models are also used to examine tertiary structure of the nanostructures and ensure that the complexes are not overly strained. Nucleic acid design has similar goals to protein design: in both, the sequence of monomers is designed to favor the desired folded or associated structure and to disfavor alternate structures. Nucleic acid design has the advantage of being a much computationally simpler problem, since the simplicity of Watson-Crick base pairing rules leads to simple heuristic methods which yield experimentally robust designs. However, nucleic acid structures are less versatile than proteins in their functionality.

Structural DNA Nanotechnology

Structural DNA nanotechnology, sometimes abbreviated as SDN, focuses on synthesizing and characterizing nucleic acid complexes and materials with various nanoscale structures. Structural DNA nanotechnology is largely based on the fact that the three-dimensional structure of DNA—the nucleic acid double helix— has a robust, defined geometry which makes it possible to predict and design the structures of more complex DNA molecules. Many such structures have been created, including two- and three-dimensional structures; and periodic, aperiodic, and discrete structures.

Periodic Lattices

Smaller nucleic acid assemblies can be equipped with sticky ends in order to combine them into a two-dimensional periodic lattice. The earliest example of this was the array of DX, or double-crossover, molecules. Each DX molecule can be designed with four sticky ends, one at each end of the two double-helical domains, and these sticky ends can be designed with sequences that cause the DX units to combine into a specific tessellated pattern. They thus form extended flat sheets which are essentially rigid two-dimensional crystals of DNA. Two-dimensional arrays have been made out of other motifs as well, including the Holliday junction rhombus array as well as various DX-based arrays making use of a double-cohesion scheme.

Creating three-dimensional lattices out of DNA was the earliest goal of DNA nanotechnology, but proved to be one of the most difficult to realize. Success in constructing three-dimensional DNA lattices was finally reported in 2009 using a motif based on the concept of tensegrity, a balance between tension and compression forces.

Nanotubes

In addition to flat sheets, DX arrays have been made to form hollow nanotubes of 4–20 nm diameter. These DNA nanotubes are somewhat similar in size and shape to carbon nanotubes, but the carbon nanotubes are stronger and better conductors, whereas the DNA nanotubes are more easily modified and connected to other structures. There have been multiple schemes for constructing DNA nanotubes, one of which uses the inherent curvature of DX tiles to form a DX lattice to curl around itself and close into a tube. An alternative design uses single-stranded "tiles" for which the rigidity of the tube is an emergent property. This method also has the benefit of being able to determine the circumference of the nanotube in a simple, modular fashion.

Polyhedra

A number of three-dimensional DNA molecules have been made which have the connectivity of a polyhedron such as an octahedron or cube. In other words, the DNA duplexes trace the edges of a polyhedron with a DNA junction at each vertex. The earliest demonstrations of DNA polyhedra involved multiple ligations and solid-phase synthesis steps to create catenated polyhedra. More recent work has yielded polyhedra whose synthesis is much easier. These include a DNA octahedron made from a long single strand designed to fold into the correct conformation, as well as a tetrahedron which can be produced from four DNA strands in a single step, pictured at the top of this article.

Arbitrary Shapes

Nanostructures of arbitrary shapes are usually made using the DNA origami method. DNA origami makes use of a long natural virus strand as a "scaffold" strand, and computationally designs shorter "staple" strands which bind to portions of the scaffold strand and force it to fold into the desired shape.

This method has the advantage of being easy to design, as the base sequence is predetermined by the scaffold strand sequence, and it also does not require high strand purity and accurate stoichiometry, as most other DNA nanotechnology methods do. DNA origami was first demonstrated for two-dimensional shapes; demonstrated designs included the smiley face and a coarse map of North America. This was later extended to solid three-dimensional shapes. In addition, structures have been constructed with two-dimensional faces which fold into an overall three-dimensional shape, akin to a cardboard box. These can be programmed to open and release their cargo in response to a stimulus, making them potentially useful as programmable molecular cages.

Functional Nucleic acid Nanostructures

DNA nanotechnology focuses on creating molecules with designed functionalities as well as structures. These include both dynamic functionality within the nucleic acid structure itself, for example with computation and mechanical motion, as well as by including other components such as small molecules or nanoparticles which have their own functionalities. Many classes of functional systems have been demonstrated.

Nanoarchitecture

The idea of using DNA arrays to template the assembly of other functional molecules was first suggested by Nadrian Seeman in 1987, but only recently has progress been made in reducing these kinds of schemes to practice. In 2006, researchers covalently attached gold nanoparticles to a DX-based tile and showed that self-assembly of the DNA structures also assembled the nanoparticles hosted on them.

Also that year, Dwyer and LaBean demonstrated the letters "D" "N" and "A" created on a 4x4 DX array using streptavidin, and a hierarchical assembly based on this approach was also demonstrated that scales to larger arrays (8X8 and 8.96 MD).

A non-covalent hosting scheme was shown in 2007, using Dervan polyamides on a DX array to arrange streptavidin proteins on specific kinds of tiles on the DNA array. There has also been interest in using DNA nanotechnology to assemble molecular electronics devices. To this end, DNA has been used to assemble single walled carbon nanotubes into field-effect transistors.

Algorithmic Self-assembly

DNA nanotechnology has been applied to the related field of DNA computing. The DX tiles can have their sticky end sequences chosen so that they act as Wang tiles, allowing them to perform computation. A DX array has been demonstrated whose assembly encodes an XOR operation; this allows the DNA array to implement a cellular automaton which generates a fractal called the Sierpinski gasket. Another system has the function of a binary counter, displaying a representation of increasing binary numbers as it grows. These results show that computation can be incorporated into the assembly of DNA arrays, increasing its scope beyond simple periodic arrays.

Figure: *The Sierpinski gasket fractal.*

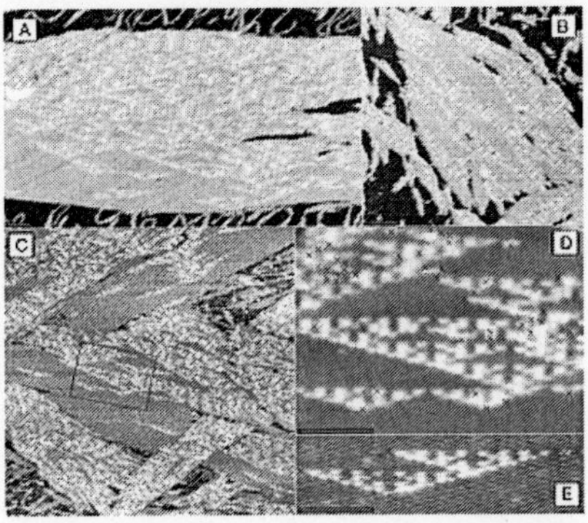

Figure: *DNA arrays that display a representation of the Sierpinski gasket on their surfaces.*

Note that DNA computing overlaps with, but is distinct from, DNA nanotechnology. The latter uses the specificity of Watson-Crick basepairing to make novel structures out of DNA. These structures can be used for DNA computing, but they do not have to be. Additionally, DNA computing can be realized without using the types of molecules made possible by DNA nanotechnology.

Nanomechanical Devices

DNA complexes have been made which change their conformation upon some stimulus. These are intended to have applications in nanorobotics. DNA machines have also been made which show a twisting motion. The first such device made use of the transition between the B-DNA and Z-DNA forms to respond to a change in buffer conditions. This reliance on buffer conditions, however, caused all devices to change state at the same time. A subsequent system, called "molecular tweezers," changes from an open to a closed state based upon the presence of control strands, allowing multiple devices to be individually operated in solution. This was followed up by another system which relies on the presence of control strands to switch from a paranemic-crossover (PX) conformation to a double-junction (JX2) conformation.

Nucleic acid nanomachines have been made which exhibit directional motion along a linear track, called DNA walkers. A large number of schemes have been demonstrated. One strategy is to control the motion of the walker along the track using control strands which need to be manually added in sequence. Another approach is to make use of restriction enzymes or deoxyribozymes to cleave the strands and cause the walker to move forward, which has the advantage of running autonomously. A later system extended the concept of DNA walkers to walk upon a two-dimensional surface rather than a linear track, and demonstrated the ability to selectively pick up and move molecular cargo. Additionally, a linear walker has been demonstrated which performs DNA-templated synthesis as the walker advances along the track, allowing autonomous multistep chemical synthesis directed by the walker.

Applications

DNA nanotechnology provides one of the only ways to form designed, complex structures with precise control over nanoscale features. The field is beginning to see application to solve basic science problems in structural biology and biophysics. One such application, the earliest envisaged for the field, is in crystallography, where molecules which are hard to crystallize by themselves could be arranged and oriented within a three-dimensional nucleic acid lattice, thus allowing determination of their structure. DNA origami rods have also been used to replace liquid crystals in residual dipolar coupling experiments in protein NMR spectroscopy; using DNA origami is advantageous because, unlike liquid crystals, they are tolerant of the detergents needed to suspend membrane proteins in solution. It has been demonstrated that DNA walkers can be used as nanoscale assembly lines to move nanoparticles and direct chemical synthesis. Furthermore, DNA origami structures have aided in the biophysical studies of enzyme function and protein folding.

DNA nanotechnology is also moving towards potential real-world applications. It has been suggested that the ability of nucleic acid arrays to arrange other molecules has potential applications in molecular scale electronics, with the assembly of a nucleic acid lattice templating the assembly of molecular electronic elements such as molecular wires. DNA nanotechnology has also been called a form of programmable matter because of the coupling of computation to its material properties. There are also potential applications for DNA nanotechnology in nanomedicine, making use of its ability to perform computation in a biocompatible format to make “smart drugs” for targeted drug delivery. One such system being investigated uses a hollow DNA box containing proteins which induce apoptosis, or cell death, which will only open when in proximity to a cancer cell.

Materials and Methods

The sequences of the individual DNA strands which make up the target structure are designed computationally, using molecular modeling and thermodynamic modeling software. Once the sequences have been designed, the nucleic acid molecules themselves are synthesized through standard oligonucleotide synthesis methods. This process is usually automated by using a machine called an oligonucleotide synthesizer, and nucleic acids of custom sequence are commercially available from many vendors. For methods which require pure strands of known concentration, the nucleic acid strands can be purified using denaturing gel electrophoresis, and concentrations are determined by one of several nucleic acid quantitation methods using ultraviolet absorbance spectroscopy. The fully formed target structures are usually characterized by native gel electrophoresis, which provides information about the size and shape of DNA molecules. An electrophoretic mobility shift assay can indicate whether a stucture incorporates all the individual desired strands. Fluorescent labeling and Förster resonance energy transfer (FRET) are also used to characterize the structure of the molecules.

Nucleic acid structures can be directly imaged by atomic force microscopy, which is well-suited to extended two-dimensional structures, but is less useful for discrete three-dimensional structures due to the interaction of the microscope tip with the fragile nucleic acid structure. For these latter structures transmission electron microscopy and cryo-electron microscopy are important methods. Extended three-dimensional lattices are analyzed by X-ray crystallography.

History

The concept of DNA nanotechnology was invented by Nadrian Seeman in the early 1980s. Seeman was originally concerned with using a three-dimensional DNA lattice to orient target molecules, which would simplify their crystallographic study by eliminating the difficult process of obtaining pure crystals. This idea had reportedly come to him in fall 1980, after realizing the similarity between the M. C. Escher woodcut Depth and an array of DNA six-arm junctions.

To this end, Seeman's laboratory in 1991 published the synthesis of a cube made of DNA, the first three-dimensional nanoscale object, for which he received the 1995 Feynman Prize in Nanotechnology, which was followed by a DNA truncated octahedron. However, it soon became clear that these molecules, polygonal shapes with flexible junctions as their vertices, were not rigid enough to form extended three-dimensional lattices. Seeman developed the more rigid double-crossover (DX) motif, and in collaboration with Erik Winfree, in 1998 published the creation of two-dimensional lattices of DX tiles. The synthesis of a three-dimensional lattice was finally published by Seeman in 2009, nearly thirty years after he had set out to do so.

These tile-based structures had the advantage that they provided the capability to implement DNA computing, which was demonstrated by Winfree and Paul Rothemund in their 2004 paper on the algorithmic self-assembly of a Sierpinski gasket structure, and for which they shared the 2006 Feynman Prize in Nanotechnology. Winfree's key insight was that the DX tiles could be used as Wang tiles, meaning that their assembly was capable of performing computation.

The first DNA nanomachine—a motif which changes its structure in response to an input—was demonstrated in 1999 by Seeman. An improved system was demonstrated by Bernard Yurke the following year, which was also the first nucleic acid device to make use of toehold-mediated strand displacement. The next advance was to translate this into mechanical motion, and in 2004 and 2005, a number of DNA walker systems were demonstrated by the groups of Seeman, Niles Pierce, Andrew Turberfield, and Chengde Mao.

In 2006, Rothemund first demonstrated the new DNA origami technique for easily and robustly creating folded DNA molecules of any shape. Rothemund had conceived of this method as being conceptually intermediate between Seeman's DX lattices, which used many short strands, and William Shih's DNA octahedron, which consisted mostly of one very long strand; Rothemund's DNA origami contains a long strand whose folding is assisted by a number of short strands. This method allowed the creation of much larger structures than were previously possible, which are less technically demanding to design and synthesize. DNA origami was the cover story of Nature on March 15, 2006. Rothemund's research demonstrating two-dimensional DNA origami structures was followed by the demonstration of solid three-dimensional DNA origami by Shih in 2009, while the labs of Jørgen Kjems and Hao Yan demonstrated hollow three-dimensional structures made out of two-dimensional faces.

DNA nanotechnology as a field was initially met with some skepticism due to the unusual non-biological use of nucleic acids as a material for building structures and doing computation, as well as the preponderance of proof of principle experiments which extended the capabilities of the field but were far from actual applications. Seeman's seminal 1991 paper on the synthesis of the DNA cube was rejected by the journal Science after one reviewer praised its originality while another criticized it for its lack of biological relevance. By the early 2010s, however, the field was considered to have increased its capabilities to the point that applications for basic science research were beginning to be realized, and practical applications in medicine and other fields were beginning to be considered feasible. In addition, the field had grown from very few active laboratories in 2001, to at least 60 in 2010, which increased the talent pool and thus the number of advancements in the field during that decade.

History and Anthropology

Because DNA collects mutations over time, which are then inherited, it contains historical information, and, by comparing DNA sequences, geneticists can infer the evolutionary history of organisms, their phylogeny. This field of phylogenetics is a powerful tool in evolutionary biology. If DNA sequences within a species are compared, population geneticists can learn the history of particular populations. This can be used in studies ranging from ecological genetics to anthropology; For example, DNA evidence is being used to try to identify the Ten Lost Tribes of Israel. DNA has also been used to look at modern family relationships, such as establishing family relationships between the descendants of Sally Hemings and Thomas Jefferson. This usage is closely related to the use of DNA in criminal investigations detailed above. Indeed, some criminal investigations have been solved when DNA from crime scenes has matched relatives of the guilty individual.

History of DNA Research

DNA was first isolated by the Swiss physician Friedrich Miescher who, in 1869, discovered a microscopic substance in the pus of discarded surgical bandages. As it resided in the nuclei of cells, he called it "nuclein". In 1878, Albrecht Kossel isolated the non-protein component of "nuclein", nucleic acid, and later isolated its five primary nucleobases. In 1919, Phoebus Levene identified the base, sugar and phosphate nucleotide unit. Levene suggested that DNA consisted of a string of nucleotide units linked together through the phosphate groups.

However, Levene thought the chain was short and the bases repeated in a fixed order. In 1937 William Astbury produced the first X-ray diffraction patterns that showed that DNA had a regular structure.

In 1927 Nikolai Koltsov proposed that inherited traits would be inherited via a "giant hereditary molecule" which would be made up of "two mirror strands that would replicate in a semi-conservative fashion using each strand as a template". In 1928, Frederick Griffith discovered that traits of the "smooth" form of the Pneumococcus could be transferred to the "rough" form of the same bacteria by mixing killed "smooth" bacteria with the live "rough" form. This system provided the first clear suggestion that DNA carries genetic information—the Avery–MacLeod–McCarty experiment—when Oswald Avery, along with coworkers Colin MacLeod and Maclyn McCarty, identified DNA as the transforming principle in 1943. DNA's role in heredity was confirmed in 1952, when Alfred Hershey and Martha Chase in the Hershey–Chase experiment showed that DNA is the genetic material of the T2 phage.

In 1953, James D. Watson and Francis Crick suggested what is now accepted as the first correct double-helix model of DNA structure in the journal Nature. Their double-helix, molecular model of DNA was then based on a single X-ray diffraction image (labeled as "Photo 51") taken by Rosalind Franklin and Raymond Gosling in May 1952, as well as the information that the DNA bases are paired — also obtained through private communications from Erwin Chargaff in the previous years. Chargaff's rules played a very important role in establishing double-helix configurations for B-DNA as well as A-DNA.

Experimental evidence supporting the Watson and Crick model were published in a series of five articles in the same issue of Nature. Of these, Franklin and Gosling's paper was the first publication of their own X-ray diffraction data and original analysis method that partially supported the Watson and Crick model; this issue also contained an article on DNA structure by Maurice Wilkins and two of his colleagues, whose analysis and in vivo B-DNA X-ray patterns also supported the presence in vivo of the double-helical DNA configurations as proposed by Crick and Watson for their double-helix molecular model of

DNA in the previous two pages of Nature. In 1962, after Franklin's death, Watson, Crick, and Wilkins jointly received the Nobel Prize in Physiology or Medicine. However, Nobel rules of the time allowed only living recipients, but a vigorous debate continues on who should receive credit for the discovery.

In an influential presentation in 1957, Crick laid out the central dogma of molecular biology, which foretold the relationship between DNA, RNA, and proteins, and articulated the "adaptor hypothesis". Final confirmation of the replication mechanism that was implied by the double-helical structure followed in 1958 through the Meselson–Stahl experiment. Further work by Crick and coworkers showed that the genetic code was based on non-overlapping triplets of bases, called codons, allowing Har Gobind Khorana, Robert W. Holley and Marshall Warren Nirenberg to decipher the genetic code. These findings represent the birth of molecular biology.

Crystallography

Crystallography is the experimental science of the arrangement of atoms in solids. The word "crystallography" derives from the Greek words crystallon = cold drop / frozen drop, with its meaning extending to all solids with some degree of transparency, and grapho = write. Before the development of X-ray diffraction crystallography, the study of crystals was based on their geometry. This involves measuring the angles of crystal faces relative to theoretical reference axes (crystallographic axes), and establishing the symmetry of the crystal in question. The former is carried out using a goniometer. The position in 3D space of each crystal face is plotted on a stereographic net, e.g. Wulff net or Lambert net. In fact, the pole to each face is plotted on the net. Each point is labelled with its Miller index. The final plot allows the symmetry of the crystal to be established.

Crystallographic methods now depend on the analysis of the diffraction patterns of a sample targeted by a beam of some type. Although X-rays are most commonly used, the beam is not always electromagnetic radiation. For some purposes electrons or neutrons are used. This is facilitated by the wave properties of the particles. Crystallographers often explicitly state the type of illumination used when referring to a method, as with the terms X-ray diffraction, neutron diffraction and electron diffraction. These three types of radiation interact with the specimen in different ways. X-rays interact with the spatial distribution of the valence electrons, while electrons are charged particles and therefore feel the total charge distribution of both the atomic nuclei and the surrounding electrons. Neutrons are scattered by the atomic nuclei through the strong nuclear forces, but in addition, the magnetic moment of neutrons is non-zero. They are therefore also scattered by magnetic fields. When neutrons are scattered from hydrogen-containing materials, they produce diffraction patterns with high noise levels. However, the material can sometimes be treated to substitute hydrogen for deuterium. Because of these different forms of interaction, the three types of radiation are suitable for different crystallographic studies.

Theory

Generally, an image of a small object is made using a lens to focus the illuminating radiation, as is done with the rays of the visible spectrum in light microscopy. However, the wavelength of visible light (about 4000 to 7000 angstroms) is three orders of magnitude longer than the length of typical atomic bonds and atoms themselves (about 1 to 2 angstroms). Therefore, obtaining information about the spatial arrangement of atoms requires the use of radiation with shorter wavelengths, such as X-rays or neutrons beam. Employing shorter wavelengths implied abandoning microscopy and true imaging, however, because there exists no material from which a lens capable of focusing this type of radiation can be created. (That said, scientists have had some success focusing X-rays with microscopic Fresnel zone plates made from gold, and by critical-angle reflection inside long tapered capillaries.) Diffracted x-ray or neutrons beams cannot be focused to produce images, so the sample structure must be reconstructed from the diffraction pattern.

Sharp features in the diffraction pattern arise from periodic, repeating structure in the sample, which are often very strong due to coherent reflection of many photons from many regularly spaced instances of similar structure, while non-periodic components of the structure result in diffuse (and usually weak) diffraction features. Said more simply, areas with a higher density and repetition of atom order tend to reflect more light toward one point in space when compared to those areas with fewer atoms and less repetition. Because of their highly ordered and repetitive structure, crystals give diffraction patterns of sharp Bragg reflection spots, and are ideal for analyzing the structure of solids.

Notation

- Coordinates in square brackets such as [100] denote a direction vector (in real space).
- Coordinates in angle brackets or chevrons such as <100> denote a family of directions which are related by symmetry operations. In the cubic crystal system for example, <100> would mean [100], [010], [001] or the negative of any of those directions.
- Miller indices in parentheses such as (100) denote a plane of the crystal structure, and regular repetitions of that plane with a particular spacing. In the cubic system, the normal to the (hkl) plane is the direction [hkl], but in lower-symmetry cases, the normal to (hkl) is not parallel to [hkl].
- Indices in curly brackets or braces such as {100} denote a family of planes and their normals which are equivalent in cubic materials due to symmetry operations, much the way angle brackets denote a family of directions. In non-cubic materials, <hkl> is not necessarily perpendicular to {hkl}.

Technique

Some materials studied using crystallography, proteins for example, do not occur naturally as crystals. Typically, such molecules are placed in solution and allowed to crystallize over days, weeks, or months through vapor diffusion. A drop of solution containing the molecule, buffer, and precipitants is sealed in a container with a reservoir containing a hygroscopic solution.

Water in the drop diffuses to the reservoir, slowly increasing the concentration and allowing a crystal to form. If the concentration were to rise more quickly, the molecule would simply precipitate out of solution, resulting in disorderly granules rather than an orderly and hence usable crystal. Once a crystal is obtained, data can be collected using a beam of radiation. Although many universities that engage in crystallographic research have their own X-ray producing equipment, synchrotrons are often used as X-ray sources, because of the purer and more complete patterns such sources can generate. Synchrotron sources also have a much higher intensity of X-ray beams, so data collection takes a fraction of the time normally necessary at weaker sources. Complementary neutron crystallography techniques are used to enhance hydrogen atoms positions. Such techniques are available in Neutron facilities. Producing an image from a diffraction pattern requires sophisticated mathematics and often an iterative process of modelling and refinement. In this process, the mathematically predicted diffraction patterns of an hypothesized or "model" structure are compared to the actual pattern generated by the crystalline sample.

Ideally, researchers make several initial guesses, which through refinement all converge on the same answer. Models are refined until their predicted patterns match to as great a degree as can be achieved without radical revision of the model. This is a painstaking process, made much easier today by computers. The mathematical methods for the analysis of diffraction data only apply to patterns, which in turn result only when waves diffract from orderly arrays. Hence crystallography applies for the most part only to crystals, or to molecules which can be coaxed to crystallize for the sake of measurement. In spite of this, a certain amount of molecular information can be deduced from the patterns that are generated by fibers and powders, which while not as perfect as a solid crystal, may exhibit a degree of order. This level of order can be sufficient to deduce the structure of simple molecules, or to determine the coarse features of more complicated molecules. For example, the double-helical structure of DNA was deduced from an X-ray diffraction pattern that had been generated by a fibrous sample.

Crystallography in Materials Engineering

Crystallography is a tool that is often employed by materials scientists. In single crystals, the effects of the crystalline arrangement of atoms is often easy to see macroscopically, because the natural shapes of crystals reflect the atomic structure.

In addition, physical properties are often controlled by crystalline defects. The understanding of crystal structures is an important prerequisite for understanding crystallographic defects. Mostly, materials do not occur in a single crystalline, but poly-crystalline form, such that the powder diffraction method plays a most important role in structural determination. A number of other physical properties are linked to crystallography. For example, the minerals in clay form small, flat, platelike structures. Clay can be easily deformed because the platelike particles can slip along each other in the plane of the plates, yet remain strongly connected in the direction perpendicular to the plates. Such mechanisms can be studied by crystallographic texture measurements.

In another example, iron transforms from a body-centered cubic (bcc) structure to a face-centered cubic (fcc) structure called austenite when it is heated. The fcc structure is a close-packed structure, and the bcc structure is not, which explains why the volume of the iron decreases when this transformation occurs. Crystallography is useful in phase identification. When performing any process on a material, it may be desired to find out what compounds and what phases are present in the material. Each phase has a characteristic arrangement of atoms. Techniques like X-ray or neutron diffraction can be used to identify which patterns are present in the material, and thus which compounds are present. Crystallography covers the enumeration of the symmetry patterns which can be formed by atoms in a crystal and for this reason has a relation to group theory and geometry.

Biology

X-ray crystallography is the primary method for determining the molecular conformations of biological macromolecules, particularly protein and nucleic acids such as DNA and RNA. In fact, the double-helical structure of DNA was deduced from crystallographic data. The first crystal structure of a macromolecule was solved in 1958 A three-dimensional model of the myoglobin molecule obtained by X-ray analysis. The Protein Data Bank (PDB) is a freely accessible repository for the structures of proteins and other biological macromolecules. Computer programs like RasMol or Pymol can be used to visualize biological molecular structures.

Neutron crystallography is often used to help refine structures obtained by x-ray methods or to solve a specific bond; the methods are often viewed as complementary, as x-rays are sensitive to electron positions and scatter most strongly off heavy atoms, while neutrons are sensitive to nucleus positions and scatter strongly off many light isotopes, including hydrogen and deuterium. Electron crystallography has been used to determine some protein structures, most notably membrane proteins and viral capsids.

DNA Error

The Singularity is Near

The Singularity Is Near: When Humans Transcend Biology is a 2005 update of Raymond Kurzweil's 1999 book, The Age of Spiritual Machines and his 1990 book The Age of Intelligent Machines. In it, as in the two previous versions, Kurzweil attempts to give a glimpse of what awaits us in the near future.

He proposes a coming technological singularity, and how we would thus be able to augment our bodies and minds with technology. He describes the singularity as resulting from a combination of three important technologies of the 21st century: genetics, nanotechnology, and robotics (including artificial intelligence).

Four central postulates of the book are as follows:

1. A technological-evolutionary point known as "the singularity" exists as an achievable goal for humanity.
2. Through a law of accelerating returns, technology is progressing toward the singularity at an exponential rate.
3. The functionality of the human brain is quantifiable in terms of technology that we can build in the near future.
4. Medical advancements make it possible for a significant number of his generation (Baby Boomers) to live long enough for the exponential growth of technology to intersect and surpass the processing of the human brain. Kurzweil's speculative reasoning and selective use of growth indicators has been heavily debated and challenged. In response to this, in the last chapter he gives responses to some of the criticisms he has received.

Chapter 10

Gene Transcription

Cell signalling is part of a complex system of communication that governs basic cellular activities and coordinates cell actions. The ability of cells to perceive and correctly respond to their microenvironment is the basis of development, tissue repair, and immunity as well as normal tissue homeostasis. Errors in cellular information processing are responsible for diseases such as cancer, autoimmunity, and diabetes. By understanding cell signalling, diseases may be treated effectively and, theoretically, artificial tissues may be created. Traditional work in biology has focused on studying individual parts of cell signalling pathways. Systems biology research helps us to understand the underlying structure of cell signalling networks and how changes in these networks may affect the transmission and flow of information. Such networks are complex systems in their organization and may exhibit a number of emergent properties including bistability and ultrasensitivity. Analysis of cell signalling networks requires a combination of experimental and theoretical approaches including the development and analysis of simulations and modelling.

Unicellular and Multicellular Organism Cell Signalling

Cell signalling has been most extensively studied in the context of human diseases and signalling between cells of a single organism. However, cell signalling may also occur between the cells of two different organisms. In many mammals, early embryo cells exchange signals with cells of the uterus. In the human gastrointestinal tract, bacteria exchange signals with each other and with human epithelial and immune system cells. For the yeast *Saccharomyces cerevisiae* during mating, some cells send a peptide signal (mating factor *pheromones*) into their environment. The mating factor peptide may bind to a cell surface receptor on other yeast cells and induce them to prepare for mating.

Types of Signals

Cells communicate with each other via direct contact (juxtacrine signalling), over short distances (paracrine signalling), or over large distances and/or scales (endocrine signalling). Some cell-to-cell communication requires direct cell-cell contact. Some cells can form gap junctions that connect their cytoplasm to the cytoplasm of adjacent cells. In cardiac muscle, gap junctions between adjacent cells allows for action potential propagation from the cardiac pacemaker region of the heart to spread and coordinately cause contraction of the heart.

The Notch signalling mechanism is an example of juxtacrine signalling (also known as contact dependent signalling) in which two adjacent cells must make physical contact in order to communicate. This requirement for direct contact allows for very precise control of cell differentiation during embryonic development.

In the worm *Caenorhabditis elegans*, two cells of the developing gonad each have an equal chance of terminally differentiating or becoming a uterine precursor cell that continues to divide. The choice of which cell continues to divide is controlled by competition of cell surface signals. One cell will happen to produce more of a cell surface protein that activates the Notch receptor on the adjacent cell. This activates a feedback loop or system that reduces Notch expression in the cell that will differentiate and increases Notch on the surface of the cell that continues as a stem cell.

Many cell signals are carried by molecules that are released by one cell and move to make contact with another cell. *Endocrine* signals are called hormones. Hormones are produced by endocrine cells and they travel through the blood to reach all parts of the body. Specificity of signalling can be controlled if only some cells can respond to a particular hormone. *Paracrine* signals such as retinoic acid target only cells in the vicinity of the emitting cell. Neurotransmitters represent another example of a paracrine signal. Some signalling molecules can function as both a hormone and a neurotransmitter. For example, epinephrine and norepinephrine can function as hormones when released from the adrenal gland and are transported to the heart by way of the blood stream. Norepinephrine can also be produced by neurons to function as a neurotransmitter within the brain. Estrogen can be released by the ovary and function as a hormone or act locally via paracrine or autocrine signalling. Active species of oxygen and nitric oxide can also act as cellular messengers. This process is dubbed redox signalling.

Receptors for Cell Signals

Cells receive information from their environment through a class of proteins known as receptors. Notch is a cell surface protein that functions as a receptor. Animals have a small set of genes that code for signalling proteins that interact specifically with Notch receptors and stimulate a response in cells that express Notch on their surface. Molecules that activate (or, in some cases, inhibit) receptors can be classified as hormones, neurotransmitters, cytokines, growth factors but all of these are called receptor ligands. The details of ligand-receptor interactions are fundamental to cell signalling.

Notch acts as a receptor for ligands that are expressed on adjacent cells. While many receptors are cell surface proteins, some are found inside cells. For example, oestrogen is a hydrophobic molecule that can pass through the lipid bilayer of cell surface membranes. Oestrogen receptors inside cells of the uterus can be activated by oestrogen that comes from the ovaries, enters the target cells, and binds to oestrogen receptors. A number of transmembrane receptors for molecules that include peptide hormones and of intracellular receptors for steroid hormones exist, giving to a cell the ability to respond to a great number of hormonal and pharmacological stimuli. In diseases, often, proteins that interact with receptors are aberrantly activated, resulting in constitutively activated downstream signals.

For several types of intercellular signalling molecules that are unable to permeate the hydrophobic cell membrane due to their hydrophilic nature, the target receptor is expressed on the membrane. When such signalling molecule activates its receptor, the signal is carried into the cell usually by means of a second messenger such as cAMP.

Signalling Pathways

In some cases, receptor activation caused by ligand binding to a receptor is directly coupled to the cell's response to the ligand. For example, the neurotransmitter GABA can activate a cell surface receptor that is part of an ion channel. GABA binding to a GABA A receptor on a neuron opens a chloride-selective ion channel that is part of the receptor.

GABA A receptor activation allows negatively-charged chloride ions to move into the neuron, which inhibits the ability of the neuron to produce action potentials. However, for many cell surface receptors, ligand-receptor interactions are not directly linked to the cell's response. The activated receptor must first interact with other proteins inside the cell before the ultimate physiological effect of the ligand on the cell's behaviour is produced. Often, the behaviour of a chain of several interacting cell proteins is altered following receptor activation. The entire set of cell changes induced by receptor activation is called a signal transduction mechanism or pathway. In the case of Notch-mediated signalling, the signal transduction mechanism can be relatively simple. Activation of Notch can cause the Notch protein to be altered by a protease. Part of the Notch protein is released from the cell surface membrane and can act to change the pattern of gene transcription in the cell nucleus.

This causes the responding cell to make different proteins, resulting in an altered pattern of cell behaviour. Cell signalling research involves studying the spatial and temporal dynamics of both receptors and the components of signalling pathways that are activated by receptors in various cell types. This pathway involves changes of protein-protein interactions inside the cell, induced by an external signal. Many growth factors bind to receptors at the cell surface and stimulate cells to progress through the cell cycle and divide. Several of these receptors are kinases that start to phosphorylate themselves and other proteins when binding to a ligand.

This phosphorylation can generate a binding site for a different protein and thus induce protein-protein interaction. The ligand (called epidermal growth factor (EGF)) binds to the receptor (called EGFR). This activates the receptor to phosphorylate itself. The phosphorylated receptor binds to an adaptor protein (GRB2), which couples the signal to further downstream signalling processes.

For example, one of the signal transduction pathways that are activated is called the mitogen-activated protein kinase (MAPK) pathway. The signal transduction component labelled as "MAPK" in the pathway was originally called "ERK," so the pathway is called the MAPK/ERK pathway. The MAPK protein is an enzyme, a protein kinase that can attach phosphate to target proteins such as the transcription factor MYC and, thus, alter gene transcription and, ultimately, cell cycle progression. Many cellular proteins are activated downstream of the growth factor receptors (such as EGFR) that initiate this signal transduction pathway.

Some signalling transduction pathways respond differently depending on the amount of signalling received by the cell. For instance, the hedgehog protein activates different genes, depending on the amount of hedgehog protein present. Complex multi-component signal transduction pathways provide opportunities for feedback, signal amplification, and interactions inside one cell between multiple signals and signalling pathways.

Classification of Intercellular Communication

Within endocrinology (the study of intercellular signalling in animals) and the endocrine system, intercellular signalling is subdivided into the following classifications:

- *Intracrine* signals are produced within the target cell.
- *Autocrine* signals target the cell its self. Sometimes autocrine cells can target cells close by if they are the same type of cell as the emitting cell. An example of this is immune cells.
- *Juxtacrine* signals target adjacent (touching) cells. These signals are transmitted along cell membranes via protein or lipid components integral to the membrane and are capable of affecting either the emitting cell or cells immediately adjacent.
- *Paracrine* signals target cells in the vicinity of the emitting cell. Neurotransmitters represent an example.

- *Endocrine* signals target distant cells. Endocrine cells produce hormones that travel through the blood to reach all parts of the body.

Designing Designs for Research

Reprinted from Trochim, W. and Land, D. (1982). Designing Designs for Research. The Researcher, 1, 1, 1-6. Much contemporary social research is devoted to examining whether a program, treatment, or manipulation causes some outcome or result. For example, we might wish to know whether a new educational program causes subsequent achievement score gains, whether a special work release program for prisoners causes lower recidivism rates, whether a novel drug causes a reduction in symptoms, and so on. Cook and Campbell (1979) argue that three conditions must be met before we can infer that such a cause-effect relation exists:

1. Covariation. Changes in the presumed cause must be related to changes in the presumed effect. Thus, if we introduce, remove, or change the level of a treatment or program, we should observe some change in the outcome measures.
2. Temporal Precedence. The presumed cause must occur prior to the presumed effect.
3. No Plausible Alternative Explanations. The presumed cause must be the only reasonable explanation for changes in the outcome measures. If there are other factors which could be responsible for changes in the outcome measures we cannot be confident that the presumed cause-effect relationship is correct.

In most social research the third condition is the most difficult to meet. Any number of factors other than the treatment or program could cause changes in outcome measures. Campbell and Stanley (1966) and later, Cook and Campbell (1979) list a number of common plausible alternative explanations (or, threats to internal validity).

For example, it may be that some historical event which occurs at the same time that the program or treatment is instituted was responsible for the change in the outcome measures; or, changes in record keeping or measurement systems which occur at the same time as the program might be falsely attributed to the program. The reader is referred to standard research methods texts for more detailed discussions of threats to validity.

This paper is primarily heuristic in purpose. Standard social science methodology textbooks (Cook and Campbell 1979; Judd and Kenny, 1981) typically present an array of research designs and the alternative explanations which these designs rule out or minimize. This tends to foster a "cookbook" approach to research design - an emphasis on the selection of an available design rather than on the construction of an appropriate research strategy.

While standard designs may sometimes fit real-life situations, it will often be necessary to "tailor" a research design to minimize specific threats to validity. Furthermore, even if standard textbook designs are used, an understanding of the logic of design construction in general will improve the comprehension of these standard approaches. This paper takes a structural approach to research design. While this is by no means the only strategy for constructing research designs, it helps to clarify some of the basic principles of design logic.

Minimizing Threats to Validity

Good research designs minimize the plausible alternative explanations for the hypothesized cause-effect relationship. But such explanations may be ruled out or minimized in a number of ways other than by design. The discussion which follows outlines five ways to minimize threats to validity, one of which is by research design:

By Argument: The most straightforward way to rule out a potential threat to validity is to simply argue that the threat in question is not a reasonable one. Such an argument may be made either *a priori* or *a posteriori*, although the former will usually be more convincing than the latter. For example, depending on the situation, one might argue that an instrumentation threat is not likely because the same test is used for pre and post test measurements and did not involve observers who might improve, or other such factors. In most cases, ruling out a potential threat to validity by argument alone will be weaker than the other approaches listed below.

As a result, the most plausible threats in a study should not, except in unusual cases, be ruled out by argument only.

By Measurement or Observation: In some cases it will be possible to rule out a threat by measuring it and demonstrating that either it does not occur at all or occurs so minimally as to not be a strong alternative explanation for the cause-effect relationship. Consider, for example, a study of the effects of an advertising campaign on subsequent sales of a particular product. In such a study, history (i.e., the occurrence of other events which might lead to an increased desire to purchase the product) would be a plausible alternative explanation. For example, a change in the local economy, the removal of a competing product from the market, or similar events could cause an increase in product sales. One might attempt to minimize such threats by measuring local economic indicators and the availability and sales of competing products. If there is no change in these measures coincident with the onset of the advertising campaign, these threats would be considerably minimized. Similarly, if one is studying the effects of special mathematics training on math achievement scores of children, it might be useful to observe everyday classroom behaviour in order to verify that students were not receiving any additional math training to that provided in the study.

By Design: Here, the major emphasis is on ruling out alternative explanations by adding treatment or control groups, waves of measurement, and the like. This topic will be discussed in more detail below.

By Analysis: There are a number of ways to rule out alternative explanations using statistical analysis. One interesting example is provided by Jurs and Glass (1971). They suggest that one could study the plausibility of an attrition or mortality threat by conducting a two-way analysis of variance.

One factor in this study would be the original treatment group designations (i.e., program vs. comparison group), while the other factor would be attrition (i.e., dropout vs. non-dropout group). The dependent measure could be the pretest or other available pre-program measures. A main effect on the attrition factor would be indicative of a threat to external validity or generalizability, while an interaction between group and attrition factors would point to a possible threat to internal validity. Where both effects occur, it is reasonable to infer that there is a threat to both internal and external validity.

The plausibility of alternative explanations might also be minimized using covariance analysis. For example, in a study of the effects of "workfare" programs on social welfare case loads, one plausible alternative explanation might be the status of local economic conditions. Here, it might be possible to construct a measure of economic conditions and include that measure as a covariate in the statistical analysis. One must be careful when using covariance adjustments of this type — "perfect" covariates do not exist in most social research and the use of imperfect covariates will not completely adjust for potential alternative explanations. Nevertheless causal assertions are likely to be strengthened by demonstrating that treatment effects occur even after adjusting on a number of good covariates.

By Preventive Action: When potential threats are anticipated they can often be ruled out by some type of preventive action. For example, if the program is a desirable one, it is likely that the comparison group would feel jealous or demoralized. Several actions can be taken to minimize the effects of these attitudes including offering the program to the comparison group upon completion of the study or using program and comparison groups which have little opportunity for contact and communication. In addition, auditing methods and quality control can be used to track potential experimental dropouts or to insure the standardization of measurement. The five categories listed above should not be considered mutually exclusive. The inclusion of measurements designed to minimize threats to validity will obviously be related to the design structure and is likely to be a factor in the analysis. A good research plan should, where possible. Make use of multiple methods for reducing threats.

In general, reducing a particular threat by design or preventive action will probably be stronger than by using one of the other three approaches. The choice of which strategy to use for any particular threat is complex and depends at least on the cost of the strategy and on the potential seriousness of the threat.

Advances in Quasi-Experimentation

Reprinted from Trochim, W. (Ed.), (1986). *Editor's Notes.* Advances in quasi-experimental design and analysis. New Directions for Program Evaluation Series, Number 31, San Francisco, CA: Jossey-Bass. The intent of this volume is to update, perhaps even to alter, our thinking about quasi-experimentation in applied social research and program evaluation. Since Campbell and Stanley (1963) introduced the term quasi-experiment, we have tended to see this area as involving primarily two interrelated topics: the theory of the validity of casual inferences and a taxonomy of the research designs that enable us to examine causal hypotheses.

We can see this in the leading expositions of quasi-experimentation (Campbell and Stanley, 1963, 1966; Cook and Campbell, 1979) as well as in the standard textbook presentations of the topic (Kidder and Judd, 1986; Rossi and Freeman, 1985), where it is typical to have separate sections or chapters that discuss validity issues first and then proceed to distinguishable quasi-experimental designs (for example, the pretest-posttest non-equivalent group design, the regression-discontinuity design, the interrupted time series design). My first inclination in editing this volume was to emulate this tradition, beginning the volume with a chapter on validity and following it with a chapter for each of the major quasi-experimental designs that raised the relevant conceptual and analytical issues and discussed recent advances. But, I think, such an approach would have simply contributed to a persistent confusion about the nature of quasi-experimentation and its role in research. Instead, this volume makes the case that we have moved beyond the traditional thinking on quasi-experiments as a collection of specific designs and threats to validity toward a more integrated, synthetic view of quasi-experimentation as part of a general logical and epistemological framework for research.

To support this view that the notion of quasi-experimentation is evolving toward increasing integration, I will discuss a number of themes that seem to characterize our current thinking and that cut across validity typologies and design taxonomies. This list of themes may also be viewed as a tentative description of the advances in our thinking about quasi-experimentation in social research.

The Case for Tailored Designs

Early expositions of quasi-experimentation took a largely taxonomic approach, laying out a collection of relatively discrete research designs and discussing how weak or strong they were for valid causal inference. Almost certainly, early proponents recognized that there was a virtual infinity of design variations and that validity was more complexly related to theory and context than their presentations implied. Nonetheless, what seemed to evolve was a "cookbook" approach to quasi-experimentation that involved "choosing" a design that fit the situation and checking off lists of validity threats. In an important paper on the coupling of randomized and non-randomized design features, Boruch (1975) explicitly encouraged us to construct research designs as combinations of more elemental units (for example, assignment strategies, measurement occasions) based on the specific contextual needs and plausible alternative explanations for a treatment effect.

This move toward hybrid, tailored, or patched-up designs, which involved suggesting how such designs could be accomplished, is one in which I have been a minor participant (Trochim and Land, 1982; Trochim, 1984). It is emphasized by Cordray in Chapter One of this volume. The implication for current practice is that we should focus on the advantages of different combinations of design features rather than on a relatively restricted set of prefabricated designs.

In teaching quasi-experimental methods, we need to break away from a taxonomic design mentality and emphasize design principles and issues that cut across the traditional distinctions between true experiments, non-experiments, and quasi-experiments.

Attention to Program Implementation

A theory-driven approach to quasi-experimentation will be futile unless we can demonstrate that the program was in fact carried out or implemented as the theory intended. Consequently, we have seen the development of program implementation theory (for example, McLaughlin, 1984) that directly addresses the process of program execution.

One approach emphasizes the development of organizational procedures and training systems that accurately transmit the program and that anticipate likely institutional sources of resistance. Another strategy involves the assessment of program delivery through program audits, management information systems, and the like. This emphasis on program implementation has further obscured the traditional distinction between process and outcome evaluation. At the least, it is certainly clear that good quasi-experimental outcome evaluation cannot be accomplished without attending to program processes, and we are continuing to develop better notions of how to combine these two efforts.

The Importance of Quality Control

Over and over, our experience with quasi-experimentation has shown that even the best-laid research plans often go awry in practice, sometimes with disastrous results. Thus, over the past decade we have begun to pay increasing attention to the integrity and quality of our research methods in real-world settings. One way of achieving this goal is to incorporate techniques used by other professions — accounting, auditing, industrial quality control — that have traditions in data integrity and quality assurance (Trochim and Visco, 1985). For instance, double bookkeeping can be used to keep verifiable records of research participation.

Acceptance sampling can be an efficient method for checking accuracy in large data collection efforts, where an exhaustive examination of records is impractical or excessive in cost. These issues are particularly important in quasi-experimentation, where it is incumbent upon the researcher to demonstrate that sampling, measurement, group assignment, and analysis decisions do not interact with program participation in ways that can confound the final interpretation of results.

The Advantages of Multiple Perspectives

We have long recognized the importance of replication and systematic variation in research. In the past few years, Cook (1985) and colleagues Shadish and Houts have articulated a rationale for achieving systematic variation that they term critical multiplism.

This perspective rests on the notion that no single realization will ever be sufficient for understanding a phenomenon with validity. Multiple realizations — of research questions, measures, samples, designs, analyses, replications, and so on — are essential for convergence on the truth of a matter. However, such a varied approach can become a methodological and epistemological Pandora's box unless we apply critical judgment in deciding which multiples we will emphasize in a study or set of studies.

Development of Increasingly Complex Realistic Analytic Models

In the past decade, we have made considerable progress toward complicating our statistical analyses to account for increasingly complex contexts and designs. One such advance involves the articulation of causal models of the sort described by Reichardt and Gollob in Chapter Six, especially models that allow for latent variables and that directly model measurement error Joreskog and Sorbom, 1979.

Another important recent development involves analyses that address the problem of selection bias or group nonequivalence — a central issue in quasi-experiments because random assignment is not used and there is no assurance that comparison groups are initially equivalent. At the same time, there is increasing recognition of the implications of not attending to the correct unit of analysis when analyzing the data and of the advantages and implications of conducting analyses at multiple levels. Thus, when we assign classrooms to conditions but analyze individual student data rather than classroom aggregates, we are liable to get a different view of program effects than we are when we analyze at the classroom level, as Shadish, Cook, and Houts argue in Chapter Two. Other notable advances that are not explicitly addressed in this volume include the development of log linear, probit, and logit models for the analysis of qualitative or nominal level outcome variables (Feinberg, 1980; Forthofer and Lehnen, 1981) and the increasing proliferation of Bayesian statistical approaches to quasi-experimental contexts (Pollard, 1986).

Parallel to the development of these increasingly complex, realistic analytic models, cynicism has deepened about the ability of any single model or analysis to be sufficient. Thus, in Chapter Six Reichardt and Gollob call for multiple analyses to bracket bias, and in Chapter Five Rindskopf recognizes the assumptive notions of any analytic approach to selection bias. We have virtually abandoned the hope of a single correct analysis, and we have accordingly moved to multiple analyses that are based on systematically distinct assumptional frameworks and that rely in an increasingly direct way on the role of judgment.

All the developments just outlined point to an increasingly realistic and complicated life for quasi-experimentalists. The overall picture that emerges is that all quasi-experimentation is judgmental. It is based on multiple and varied sources of evidence, it should be multiplistic in realization, it must attend to process as well as to outcome, it is better off when theory driven, and it leads ultimately to multiple analyses that attempt to bracket the program effect within some reasonable range.

In one sense, this is hardly a pretty picture. Our views about quasi-experimentation and its role in causal inference are certainly more tentative and critical than they were in 1965 or perhaps even in 1979. But, this more integrated and complex view of quasi-experimentation has emerged directly from our experiences in the conduct of such studies. As such, it realistically represents our current thinking about one of the major strands in the evolution of social research methodology in this century.

Analysis

By the time you get to the analysis of your data, most of the really difficult work has been done. It's much more difficult to: define the research problem; develop and implement a sampling plan; conceptualize, operationalize and test your measures; and develop a design structure. If you have done this work well, the analysis of the data is usually a fairly straightforward affair. In most social research the data analysis involves three major steps, done in roughly this order:

- Cleaning and organizing the data for analysis (Data Preparation)
- Describing the data (Descriptive Statistics)
- Testing Hypotheses and Models (Inferential Statistics)

Data Preparation involves checking or logging the data in; checking the data for accuracy; entering the data into the computer; transforming the data; and developing and documenting a database structure that integrates the various measures. Descriptive Statistics are used to describe the basic features of the data in a study. They provide simple summaries about the sample and the measures. Together with simple graphics analysis, they form the basis of virtually every quantitative analysis of data. With descriptive statistics you are simply describing what is, what the data shows.

Inferential Statistics investigate questions, models and hypotheses. In many cases, the conclusions from inferential statistics extend beyond the immediate data alone. For instance, we use inferential statistics to try to infer from the sample data what the population thinks. Or, we use inferential statistics to make judgments of the probability that an observed difference between groups is a dependable one or one that might have happened by chance in this study. Thus, we use inferential statistics to make inferences from our data to more general conditions; we use descriptive statistics simply to describe what's going on in our data.

In most research studies, the analysis section follows these three phases of analysis. Descriptions of how the data were prepared tend to be brief and to focus on only the more unique aspects to your study, such as specific data transformations that are performed. The descriptive statistics that you actually look at can be voluminous. Usually, the researcher links each of the inferential analyses to specific research questions or hypotheses that were raised in the introduction, or notes any models that were tested that emerged as part of the analysis. In most analysis write-ups it's especially critical to not "miss the forest for the trees." If you present too much detail, the reader may not be able to follow the central line of the results. Often extensive analysis details are appropriately relegated to appendices, reserving only the most critical analysis summaries for the body of the report itself.

DNA Biochemistry

Deoxyribonucleic acid, or DNA, is a nucleic acid that contains the genetic instructions used in the development and functioning of all known living organisms with the exception of some viruses. The main role of DNA molecules is the long-term storage of information. DNA is often compared to a set of blueprints, like a recipe or a code, since it contains the instructions needed to construct other components of cells, such as proteins and RNA molecules. The DNA segments that carry this genetic information are called genes, but other DNA sequences have structural purposes, or are involved in regulating the use of this genetic information.

DNA consists of two long polymers of simple units called nucleotides, with backbones made of sugars and phosphate groups joined by ester bonds. These two strands run in opposite directions to each other and are therefore anti-parallel. Attached to each sugar is one of four types of molecules called bases. It is the sequence of these four bases along the backbone that encodes information. This information is read using the genetic code, which specifies the sequence of the amino acids within proteins. The code is read by copying stretches of DNA into the related nucleic acid RNA, in a process called transcription.

Within cells, DNA is organized into long structures called chromosomes. These chromosomes are duplicated before cells divide, in a process called DNA replication. Eukaryotic organisms (animals, plants, fungi, and protists) store most of their DNA inside the cell nucleus and some of their DNA in organelles, such as mitochondria or chloroplasts. In contrast, prokaryotes (bacteria and archaea) store their DNA only in the cytoplasm. Within the chromosomes, chromatin proteins such as histones compact and organize DNA. These compact structures guide the interactions between DNA and other proteins, helping control which parts of the DNA are transcribed.

Properties

DNA is a long polymer made from repeating units called nucleotides. As first discovered by James D. Watson and Francis Crick, the structure of DNA of all species comprises two helical chains each coiled round the same axis, and each with a pitch of 34 Ångströms (3.4 nanometres) and a radius of 10 Ångströms (1.0 nanometres). According to another study, when measured in a particular solution, the DNA chain measured 22 to 26 Ångströms wide (2.2 to 2.6 nanometres), and one nucleotide unit measured 3.3 Å (0.33 nm) long. Although each individual repeating unit is very small, DNA polymers can be very large molecules containing millions of nucleotides. For instance, the largest human chromosome, chromosome number 1, is approximately 220 million base pairs long. In living organisms, DNA does not usually exist as a single molecule, but instead as a pair of molecules that are held tightly together.

These two long strands entwine like vines, in the shape of a double helix. The nucleotide repeats contain both the segment of the backbone of the molecule, which holds the chain together, and a base, which interacts with the other DNA strand in the helix. A base linked to a sugar is called a nucleoside and a base linked to a sugar and one or more phosphate groups is called a nucleotide. If multiple nucleotides are linked together, as in DNA, this polymer is called a polynucleotide. The backbone of the DNA strand is made from alternating phosphate and sugar residues. The sugar in DNA is 2-deoxyribose, which is a pentose (five-carbon) sugar. The sugars are joined together by phosphate groups that form phosphodiester bonds between the third and fifth carbon atoms of adjacent sugar rings.

These asymmetric bonds mean a strand of DNA has a direction. In a double helix the direction of the nucleotides in one strand is opposite to their direction in the other strand: the strands are *antiparallel*. The asymmetric ends of DNA strands are called the 52 (*five prime*) and 32 (*three prime*) ends, with the 5' end having a terminal phosphate group and the 3' end a terminal hydroxyl group. One major difference between DNA and RNA is the sugar, with the 2-deoxyribose in DNA being replaced by the alternative pentose sugar ribose in RNA.

The DNA double helix is stabilized by hydrogen bonds between the bases attached to the two strands. The four bases found in DNA are adenine (abbreviated A), cytosine (C), guanine (G) and thymine (T). These four bases are attached to the sugar/phosphate to form the complete nucleotide. These bases are classified into two types; adenine and guanine are fused five- and six-membered heterocyclic compounds called purines, while cytosine and thymine are six-membered rings called pyrimidines. A fifth pyrimidine base, called uracil (U), usually takes the place of thymine in RNA and differs from thymine by lacking a methyl group on its ring.

Uracil is not usually found in DNA, occurring only as a breakdown product of cytosine. In addition to RNA and DNA, a large number of artificial nucleic acid analogues have also been created to study the proprieties of nucleic acids, or for use in biotechnology.

Grooves

Twin helical strands form the DNA backbone. Another double helix may be found by tracing the spaces, or grooves, between the strands. These voids are adjacent to the base pairs and may provide a binding site. As the strands are not directly opposite each other, the grooves are unequally sized.

One groove, the major groove, is 22 Å wide and the other, the minor groove, is 12 Å wide. The narrowness of the minor groove means that the edges of the bases are more accessible in the major groove. As a result, proteins like transcription factors that can bind to specific sequences in double-stranded DNA usually make contacts to the sides of the bases exposed in the major groove. This situation varies in unusual conformations of DNA within the cell, but the major and minor grooves are always named to reflect the differences in size that would be seen if the DNA is twisted back into the ordinary B form.

Base Pairing

Each type of base on one strand forms a bond with just one type of base on the other strand. This is called complementary base pairing. Here, purines form hydrogen bonds to pyrimidines, with A bonding only to T, and C bonding only to G. This arrangement of two nucleotides binding together across the double helix is called a base pair. As hydrogen bonds are not covalent, they can be broken and rejoined relatively easily. The two strands of DNA in a double helix can therefore be pulled apart like a zipper, either by a mechanical force or high temperature. As a result of this complementarity, all the information in the double-stranded sequence of a DNA helix is duplicated on each strand, which is vital in DNA replication. Indeed, this reversible and specific interaction between complementary base pairs is critical for all the functions of DNA in living organisms.

The two types of base pairs form different numbers of hydrogen bonds, AT forming two hydrogen bonds, and GC forming three hydrogen bonds. DNA with high GC-content is more stable than DNA with low GC-content, but contrary to popular belief, this is not due to the extra hydrogen bond of a GC base pair but rather the contribution of stacking interactions (hydrogen bonding merely provides specificity of the pairing, not stability). As a result, it is both the percentage of GC base pairs and the overall length of a DNA double helix that determine the strength of the association between the two strands of DNA. Long DNA helices with a high GC content have stronger-interacting strands, while short helices with high AT content have weaker-interacting strands. In biology, parts of the DNA double helix that need to separate easily, such as the TATAAT Pribnow box in some promoters, tend to have a high AT content, making the strands easier to pull apart.

In the laboratory, the strength of this interaction can be measured by finding the temperature required to break the hydrogen bonds, their melting temperature (also called T_m value). When all the base pairs in a DNA double helix melt, the strands separate and exist in solution as two entirely independent molecules. These single-stranded DNA molecules (*ssDNA*) have no single common shape, but some conformations are more stable than others.

Sense and Antisense

A DNA sequence is called "sense" if its sequence is the same as that of a messenger RNA copy that is translated into protein. The sequence on the opposite strand is called the "antisense" sequence. Both sense and antisense sequences can exist on different parts of the same strand of DNA (i.e. both strands contain both sense and antisense sequences). In both prokaryotes and eukaryotes, antisense RNA sequences are produced, but the functions of these RNAs are not entirely clear.

One proposal is that antisense RNAs are involved in regulating gene expression through RNA-RNA base pairing. A few DNA sequences in prokaryotes and eukaryotes, and more in plasmids and viruses, blur the distinction between sense and antisense strands by having overlapping genes.

In these cases, some DNA sequences do double duty, encoding one protein when read along one strand, and a second protein when read in the opposite direction along the other strand. In bacteria, this overlap may be involved in the regulation of gene transcription, while in viruses, overlapping genes increase the amount of information that can be encoded within the small viral genome.

Supercoiling

DNA can be twisted like a rope in a process called DNA supercoiling. With DNA in its "relaxed" state, a strand usually circles the axis of the double helix once every 10.4 base pairs, but if the DNA is twisted the strands become more tightly or more loosely wound. If the DNA is twisted in the direction of the helix, this is positive supercoiling, and the bases are held more tightly together. If they are twisted in the opposite direction, this is negative supercoiling, and the bases come apart more easily. In nature, most DNA has slight negative supercoiling that is introduced by enzymes called topoisomerases. These enzymes are also needed to relieve the twisting stresses introduced into DNA strands during processes such as transcription and DNA replication.

Alternate DNA Structures

DNA exists in many possible conformations that include A-DNA, B-DNA, and Z-DNA forms, although, only B-DNA and Z-DNA have been directly observed in functional organisms. The conformation that DNA adopts depends on the hydration level, DNA sequence, the amount and direction of supercoiling, chemical modifications of the bases, the type and concentration of metal ions, as well as the presence of polyamines in solution.

The first published reports of A-DNA X-ray diffraction patterns— and also B-DNA used analyses based on Patterson transforms that provided only a limited amount of structural information for oriented fibers of DNA. An alternate analysis was then proposed by Wilkins *et al.*, in 1953, for the *in vivo* B-DNA X-ray diffraction/scattering patterns of highly hydrated DNA fibers in terms of squares of Bessel functions. In the same journal, James D. Watson and Francis Crick presented their molecular modeling analysis of the DNA X-ray diffraction patterns to suggest that the structure was a double-helix.

Although the 'B-DNA form' is most common under the conditions found in cells, it is not a well-defined conformation but a family of related DNA conformations that occur at the high hydration levels present in living cells. Their corresponding X-ray diffraction and scattering patterns are characteristic of molecular paracrystals with a significant degree of disorder. Compared to B-DNA, the A-DNA form is a wider right-handed spiral, with a shallow, wide minor groove and a narrower, deeper major groove.

The A form occurs under non-physiological conditions in partially dehydrated samples of DNA, while in the cell it may be produced in hybrid pairings of DNA and RNA strands, as well as in enzyme-DNA complexes. Segments of DNA where the bases have been chemically modified by methylation may undergo a larger change in conformation and adopt the Z form. Here, the strands turn about the helical axis in a left-handed spiral, the opposite of the more common B form. These unusual structures can be recognized by specific Z-DNA binding proteins and may be involved in the regulation of transcription.

Quadruplex Structures

At the ends of the linear chromosomes are specialized regions of DNA called telomeres. The main function of these regions is to allow the cell to replicate chromosome ends using the enzyme telomerase, as the enzymes that normally replicate DNA cannot copy the extreme 32 ends of chromosomes. These specialized chromosome caps also help protect the DNA ends, and stop the DNA repair systems in the cell from treating them as damage to be corrected. In human cells, telomeres are usually lengths of single-stranded DNA containing several thousand repeats of a simple TTAGGG sequence.

These guanine-rich sequences may stabilize chromosome ends by forming structures of stacked sets of four-base units, rather than the usual base pairs found in other DNA molecules. Here, four guanine bases form a flat plate and these flat four-base units then stack on top of each other, to form a stable *G-quadruplex* structure. These structures are stabilized by hydrogen bonding between the edges of the bases and chelation of a metal ion in the centre of each four-base unit. Other structures can also be formed, with the central set of four bases coming from either a single strand folded around the bases, or several different parallel strands, each contributing one base to the central structure.

In addition to these stacked structures, telomeres also form large loop structures called telomere loops, or T-loops. Here, the single-stranded DNA curls around in a long circle stabilized by telomere-binding proteins. At the very end of the T-loop, the single-stranded telomere DNA is held onto a region of double-stranded DNA by the telomere strand disrupting the double-helical DNA and base pairing to one of the two strands. This triple-stranded structure is called a displacement loop or D-loop.

Branched DNA

In DNA fraying occurs when non-complementary regions exist at the end of an otherwise complementary double-strand of DNA. However, branched DNA can occur if a third strand of DNA is introduced and contains adjoining regions able to hybridize with the frayed regions of the pre-existing double-strand. Although the simplest example of branched DNA involves only three strands of DNA, complexes involving additional strands and multiple branches are also possible. Branched DNA can be used in nanotechnology to construct geometric shapes.

Chapter 11

Genetic Mechanisms and Molecular Biology

In addition to very frequent simple mutations, there are several mutation classes which involve sequence exchange between allelic or nonallelic sequences, often involving repeated sequences. For example, tandemly repetitive DNA is prone to deletion/insertion polymorphism whereby different alleles vary in the number of integral copies of the tandem repeat. Such variable number of tandem repeat (VNTR) polymorphisms can occur in the case of repeated units that are very short (microsatellites); intermediate (minisatellites) or large. Different genetic mechanisms can account for VNTR polymorphism depending on the size of the repeating unit. In addition, interspersed repeats can also predispose to deletions/ duplications by a variety of different genetic mechanisms.

Slipped Strand Mispairing can Cause VNTR Polymorphism at Short Tandem Repeats (Microsatellites)

There is considerable variation in the germline mutation rates at microsatellite loci, ranging from an undetectable level up to about 8×10^{-3}. Novel length alleles at (CA)/ (TG) microsatellites and at tetranucleotide marker loci are known to be formed without exchange of flanking markers. This means that they are not generated by unequal crossover. Instead, as new mutant alleles have been observed to differ by a single repeat unit from the originating parental allele (Mahtani and Willard, 1993), the most likely mechanism to explain length variation is a form of exchange of sequence information which commences by slipped strand mispairing.

This occurs when the normal pairing between the two complementary strands of a double helix is altered by staggering of the repeats on the two strands, leading to incorrect pairing of repeats. Although slipped strand mispairing can be envisaged to occur in nonreplicating DNA, replicating DNA may offer more opportunity for slippage and hence the mechanism is often also called replication slippage or polymerase slippage. In addition to mispairing between tandem repeats, slippage replication has been envisaged to generate large deletions and duplications by mispairing between noncontiguous repeats and has been suggested to be a major mechanism for DNA sequence and genome evolution. The pathogenic potential of short tandem repeats is considerable.

Large Units of Tandemly Repeated DNA are Prone to Insertion

Homologous recombination describes recombination (crossover) occurring at meiosis or, rarely, mitosis between identical or very similar DNA sequences. It usually involves breakage of *nonsister chromatids* of a pair of homologs and rejoining of the fragments to generate new recombinant strands. Sister chromatid exchange is an analogous type of sequence exchange involving breakage of individual *sister chromatids* and rejoining fragments that initially were on different chromatids of the same chromosome. Both homologous recombination and sister chromatid exchange normally involve *equal* exchanges - cleavage and rejoining of the chromatids occurs at the same position on each chromatid. As a result, the exchanges occur *between allelic sequences* and at corresponding positions within alleles. In the case of intragenic equal crossover between two alleles, a new allele can result which is a fusion gene (or hybrid gene), comprising a terminal fragment from one allele and the remaining sequence of the second allele. However, equal sister chromatid exchanges cannot normally produce genetic variation because sister chromatids have identical DNA sequences.

Unequal crossover is a form of recombination in which the crossover takes place *between nonallelic sequences* on nonsister chromatids of a pair of homologs. Often the sequences at which crossover takes place show very considerable sequence homology which presumably stabilizes mispairing of the chromosomes. Because crossover occurs between mispaired nonsister chromatids, the exchange results in a deletion on one of the participating chromatids and an insertion on the other.

The analogous exchange between sister chromatids is called unequal sister chromatid exchange. Both mechanisms occur predominantly at locations where the tandemly repeated units are moderate to large in size. In such cases, the very high degree of sequence homology between the different repeats can facilitate pairing of nonallelic repeats on nonsister chromatids or sister chromatids. If chromosome breakage and rejoining occurs while the chromatids are mispaired in this way, an insertion or deletion of an integral number of repeat units will result.

Note that such exchanges are *reciprocal*; both participating chromatids are modified, in one case resulting in an insertion, and in the other case in a complementary deletion.

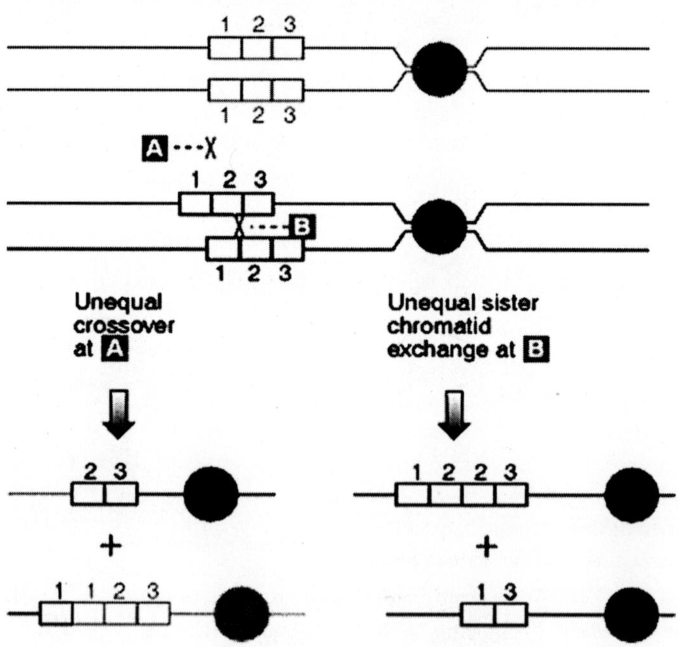

Figure: *Unequal crossover and unequal sister chromatid exchange cause insertions and deletions.*

Unequal sister chromatid exchange is thought to be a major mechanism underlying VNTR polymorphism in the rDNA clusters. Unequal crossover is also expected to occur comparatively frequently in complex satellite DNA repeats and at tandemly repeated gene loci. In the latter case, unequal crossover is known to generate pathogenic deletions at some loci. Such exchanges can also lead to concerted evolution by causing a particular variant to spread through an array of tandem repeats, resulting in *homogenisation* of the repeat units.

Occasionally, unequal crossover and unequal sister chromatid exchanges can occur at regions where there is little homology. This is likely to be the case when such mechanisms first generate a tandemly duplicated locus following mispairing of nonallelic repeats such as two *Alu* repeats or even smaller elements.

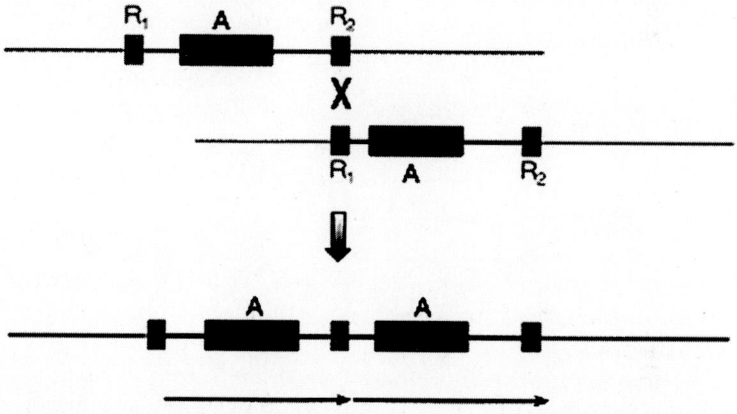

Figure: *Tandem gene duplication can result from unequal crossover or unequal sister chromatid exchange, facilitated by short interspersed repeats.*

Gene Conversion Events may be Relatively Frequent in Tandemly Repetitive DNA

Gene conversion describes a *nonreciprocal* transfer of sequence information between a pair of nonallelic DNA sequences (*interlocus gene conversion)* or allelic sequences (*interallelic gene conversion*). One of the pair of interacting sequences, the *donor*, remains unchanged. The other DNA sequence, the *acceptor*, is changed by having some or all of its sequence replaced by a sequence copied from the donor sequence.

The sequence exchange is therefore a directional one; the acceptor sequence is modified by the donor sequence, but not the other way round.

One possible mechanism for gene conversion envisages formation of a heteroduplex between a DNA strand from the donor gene and a complementary strand from the acceptor gene. Following heteroduplex formation, conversion of an acceptor gene segment may occur by mismatch repair - DNA repair enzymes recognize that the two strands of the heteroduplex are not perfectly matched and 'correct' the DNA sequence of the acceptor strand to make it perfectly complementary in the converted region to the sequence of the donor gene strand.

Gene conversion has been well-described in fungi where all four products of meiosis can be recovered and studied (tetrad analysis). In humans and mammals it is not possible to do this and so gene conversion cannot be demonstrated unambiguously in higher organisms (it can never be distinguished from double crossover events, for example, although double crossovers occurring in very close proximity would normally be expected to be extremely unlikely).

Despite the difficulty in identifying gene conversion in complex organisms, there are numerous instances in mammalian genomes where an allele at one locus shows a pattern of mutations which strongly resembles those found in alleles at another locus of the same species. Such evidence suggests gene conversion-like exchanges between loci. Although simple comparisons of two sequences may be suggestive, the evidence for gene conversion is most compelling when a new mutant allele can be compared directly with its progenitor sequence. Certain highly mutable loci lend themselves to this type of analysis. In particular, some hypervariable minisatellite loci have high germline mutation rates (often 1% or more per gamete) and individual repeats often show nucleotide differences so that repeat subclasses can be recognised.

Germline mutations can be studied by detecting and characterising mutant minisatellite alleles in individual gametes. To do this, PCR analysis has been conducted on multiple dilute aliquots of DNA isolated from the sperm of an individual (small pool PCR), where each aliquot is calibrated to contain a few, perhaps 100, input molecules (Jeffreys *et al.*, 1994). The PCR products recovered from individual pools can then be typed to identify any new mutations that result in a novel allele whose length is sufficiently different as to be distinguishable from the progenitor allele.

Analyses of the patterns of germline mutation at three such loci have failed to identify exchanges of flanking markers and have shown that most mutations occurring at these loci are polar, involving the preferential gain of a few repeats at one end of a tandem repeat array. There is a bias towards gain of repeats and evidence was obtained for nonreciprocal sequence exchange between alleles, suggesting interallelic gene conversion (Jeffreys *et al.*, 1994). Evidence for interlocus gene conversion has also been obtained in human genes, notably the steroid 21-hydroxylase gene.

Pathogenic Mutations

There is a High Deleterious Mutation Rate in Hominids

Neutral mutations (those which are neither detrimental nor advantageous for the organism carrying them) accumulate throughout the generations at a rate equal to the mutation rate. To get an estimate of the (total) mutation rate is therefore simple: all one needs to do is to measure the rate of change of some presumed neutral sequence (e.g. intronic, pseudogene, etc). The deleterious mutation rate, by contrast, has been notoriously difficult to measure and no convincing estimate existed for any vertebrate until a study reported by Eyre-Walker and Keightley (1999).

They investigated amino acid changes in 46 proteins occurring in the human ancestral line after its divergence from the chimpanzee. If all non-synonymous substitutions were neutral, 231 new substitutions would have been expected in their sample of genes (given an average neutral mutation rate of 0.0056 nonsynonymous substitutions per nucleotide and a total of 41 471 nucleotides investigated). Instead, only 143 nonsynonymous substitutions were observed and 88 such substitutions were inferred to have been removed by natural selection because they had been deleterious. On the assumption of 60 000 genes, and 240 000 generations since human-chimpanzee divergence, they estimated a deleterious rate of 1.6

mutations per person per generation out of a total of 4.2 mutations per person generation.

The estimated deleterious mutation rates in chimpanzees and gorillas were very similar but those for rodent-specific lineages are about one order of magnitude less, possibly because of much smaller numbers of germ cell divisions in rodents. The very high deleterious rate in hominids may even be an underestimate. If the total gene number were 80 000 and an average coding sequence was 1800 nucleotides, the estimated deleterious mutation rate would be 2.5 mutations per person per generation and, on other grounds too, a more likely rate has been considered to be three deleterious mutations per person per generation. So with three genetic deaths per person why are we not extinct? The data would suggest that harmful mutations need to be weeded out in clusters at a time. One way to achieve this would be if natural selection operated such that individuals with the most mutations are preferentially eliminated (e.g. harmful mutations interact). This could only happen in a sexual species where mutations are shuffled each generation by genetic recombination, and so the existence of such a high deleterious mutation rate has been taken as further vindication that sex (meiotic recombination) is an efficient way to eliminate harmful mutations.

Pathogenic Mutations are Preferentially Located at Certain Types of Intragenic DNA Sequence

Pathogenic mutations can occur at three types of DNA sequence at a gene locus.

- *The coding sequence of the gene.* This is where the great majority of recorded pathogenic mutations have been identified. Those due to nucleotide substitution are, in the vast majority of cases, nonsynonymous substitutions and mostly occur at first and second base positions of codons. However, very rarely, a synonymous codon substitution is not neutral as expected, but may cause disease by activating a cryptic splice site. Because of its relatively high mutability, the CpG dinucleotide is often located at hotspots for pathogenic mutation in coding DNA (Cooper and Youssoufian, 1988). Other hotspots include tandem repeats within coding DNA.
- *Intragenic noncoding sequences.* This is restricted to sequences which are necessary for correct expression

of the gene, such as important intronic elements, notably the highly conserved GT and AG dinucleotides at the ends of introns, but also conserved elements of the untranslated sequences. Often such mutations represent a small component (~10–15%) of the total pathogenic mutations at a gene locus (Cooper *et al.*, 1995). However, in some disorders pathogenic splicing mutations may be common. In the case of the collagen disorder osteogenesis imperfecta they constitute a very common pathological mutation which is second in frequency only to substitutions leading to replacement of the highly conserved, structurally important glycine residues. The collagen genes have small exons and a comparatively large number of introns (often more than 50 and as many as 106 in the case of the *COL7A1* gene), making them exceptional targets for splicing mutations. Occasional pathogenic mutations have been recorded in the 52 UTR (such as in the case of hemophilia B Leyden) and appear to exert their effect at the transcriptional level. Several examples are also known of pathogenic mutations in the 32 UTR.

- *Regulatory sequences outside exons.* Most mutations located in regulatory sequences have been identified in conserved elements located just upstream of the first exon, notably promoter elements. In addition, other more distantly located regulatory elements may be sites of pathological mutation. For example, deletions which eliminate the â-globin LCR but leave the â-globin gene and its promoter intact result in almost complete abolition of â-globin gene expression and contribute to â-thalassemia. Clearly, in some cases a gene may be regulated by the product of a distantly related gene.

The Mitochondrial Genome is a Hotspot for Pathogenic Mutations

Because of the very large size of the human nuclear genome, most mutations occur in nuclear DNA sequences. By comparison, the mitochondrial genome is a small target for mutation (about 1/200 000 of the size of the nuclear genome). Unlike nuclear genes, mitochondrial genes are present in numerous copies (there are thousands of copies of the mtDNA molecule in each human somatic cell; some cells, such as brain and muscle cells, have particularly high oxidative phosphorylation requirements and so more mitochondria).

The mtDNA is inherited from the maternal oocyte, which is an exceptional cell with many more mtDNA molecules than somatic cells. Given that a mutation in mitochondrial DNA must arise on a single mtDNA molecule, one might intuitively expect that the chances of a single mtDNA mutation becoming *fixed* would be very low and the mutation rate correspondingly low. On these grounds, one could anticipate that the proportion of clinical disease due to pathogenic mutation in the mitochondrial genome should be extremely low. Instead, the frequency of 'mitochondrial disorders' is rather high and the mitochondrial genome can be considered to be a mutation hotspot. Different factors can explain this apparent paradox:

- *Differential target size for pathogenic mutation.* Pathogenesis is associated with mutations in coding DNA and the mitochondrial genome has a much higher percentage of coding DNA (93%) than found in the nuclear genome (3%). When this is taken into consideration, however, there is still a large imbalance: about 100 Mb of coding DNA in the nuclear genome but only 15.4 kb of coding sequence in the mitochondrial genome, giving a target ratio of 6000:1 in favour of the nuclear genome.
- *High mutation rate in mtDNA.* The mitochondrial genome is much more prone to nucleotide change than the nuclear genome. Even although about 100 000 copies of the mitochondrial genome are maternally inherited in the fertilized oocyte there are mechanisms which permit rapid fixation of mutations in mitochondrial DNA. The combination of mtDNA instability and a high fixation rate means that the mutation rate in mitochondrial DNA is very high. Mutations have been reported to be fixed in the mitochondrial genomes of animal cells at a rate which is about 10 times greater than occurring in equivalent sequences in the nuclear genome (Brown *et al.*, 1979). This means that the small recombination-deficient animal mtDNA molecules appear to be evolving remarkably rapidly, corresponding

to about 2–4% sequence divergence per million years. In contrast, plant mtDNA molecules are comparatively large (150 kb-2.5 Mb), have introns, engage in recombination and are evolving comparatively slowly.

The high instability of mtDNA has been postulated to result from several factors. The high rate of production of reactive oxygen intermediates by the respiratory chain is thought to cause substantial oxidative damage to mtDNA which, unlike nuclear DNA, is not protected by histones. The mtDNA also has to undergo many more rounds of replication than chromosomal DNA. Although several well-characterised mtDNA repair systems are now known, some frequent mutations cannot be repaired, including thymidine dimers.

Conserved Sequence Required for Normal Splicing

Many genes naturally undergo alternative forms of RNA splicing. In addition, mutations can sometimes produce an aberrant form of RNA splicing which is pathogenic. Sometimes this results in the sequences of whole exons being excluded from the mature RNA (exon skipping) or retention of whole introns. On other occasions, the abnormal splicing pattern may exclude part of a normal exon or result in new exonic sequences. Point mutations which alter a conserved sequence that is normally required for RNA splicing are comparatively common. Occasionally, however, aberrant splicing of a gene can be induced by mutation of other sequence elements which resemble splice donor or splice acceptor sequences but which are not normally involved in splicing (cryptic splice sites).

Mutations Which Alter Important Splice Site Signals

Often such mutations occur at the essentially invariant GT and AG dinucleotides located respectively at the start of an intron (*splice donor*) or at its end (*splice acceptor*). Flanking these important signals, however, are other conserved sequence elements which, if mutated, can also cause aberrant splicing. Mutations which alter such sequences can have different consequences:

- *Failure of splicing causing intron retention.* This can occasionally result, for example, when an intron is small and the neighbouring sequence lacks alternative legitimate splice sites or *cryptic splice sites* (sequences which resemble the consensus splice site sequences but which are not normally used by the splicing apparatus). The introduction of intronic sequence into the coding sequence of a mature mRNA will, at the very least, introduce additional amino acids and may cause a frameshift.

- *Exon skipping.* The splicing apparatus uses an *alternative* legitimate splice site. Mutation of a splice donor sequence results in skipping of the upstream exon while mutation of the splice acceptor sequence results in skipping of the downstream exon. Often, the exclusion of an exon has a profound effect on gene expression: it may result in a frameshift, an unstable RNA transcript, or a nonfunctional polypeptide because of a loss of a critical group of amino acids.

In addition to the above, the *branch site* used in splicing may be mutated leading to defective splicing.

Mutations of Sequences Which are not Normally Important for RNA Splicing

Cryptic splice sites coincidentally resemble the sequences of authentic splice sites but are not normally used in splicing, unless a mutation alters the sequence so that the splicing apparatus now recognizes it as a normal splice site. The use of an intronic cryptic splice site will introduce new amino acids, while using an exonic cryptic splice site will result in a deletion of coding DNA. The former is a cautionary reminder that apparently silent mutations may yet be pathogenic. Note that in some cases mutations which occur within exons but not at cryptic splice sites can also induce skipping of that exon.

Premature Termination Codon Usually Result in Unstable mRNA but other Outcomes are Possible

Several different classes of mutation can introduce a premature termination codon (chain terminating mutations). Nonsense mutations produce a premature termination codon simply by substituting a normal codon with a stop codon. Frameshifting insertions and deletions usually also introduce a premature termination codon not too far downstream of the mutation site. This happens because there is no selection pressure to avoid stop codons in the other translational reading frames and so given established nucleotide frequencies, at least one stop codon is usually encountered within a stretch of 100 nucleotides downstream of the mutation site. A variety of splice site mutations too can introduce a premature termination codon e.g. by skipping of a single exon containing a number of nucleotides that cannot be divided by 3. There are several possible consequences for gene expression for chain-terminating mutations:

- *Unstable mRNA.* This is by far the most frequent consequence. A mRNA carrying a premature codon is usually rapidly degraded *in vivo* by a form of RNA surveillance known as nonsense-mediated mRNA decay (Hentze and Kulozik, 1999; Culbertson, 1999). This can avoid the potentially lethal consequences of producing a truncated polypeptide which could interfere with vital cell functions.
- *Truncated polypeptide.* The production of a polypeptide truncated at the C terminus is a very rare outcome *in vivo* (the well-known protein truncation test which assays for mutations introducing a premature termination codon is carried out using an *in vitro* transcription-translation system). Nevertheless, some truncated polypeptides are produced *in vivo*. The effect on gene expression may be difficult to predict and will depend among other things on the extent of the truncation, the stability of the polypeptide product and its ability to interfere with expression of normal alleles.
- *Exon skipping.* Some nonsense mutations appear to induce skipping of constitutive exons *in vivo*. For example, a nonsense mutation in the middle of exon 51 of the *FBN1* fibrillin gene (corresponding to the C terminus of the protein) causes that exon to be skipped (Dietz *et al.*, 1993). As a result of exon skipping the abnormally spliced mRNA uses the normal stop codon and escapes nonsense-mediated mRNA decay unlike any full length mRNA which may be produced from the pre-mRNA. The abnormally spliced *FBN1* mRNA accumulates and is translated to give a dominant negative protein lacking C-terminal sequences.

The Pathogenic Potential of Repeated Sequences

The human genome, like other mammalian genomes, has a very high proportion of DNA sequences that are repeated.

Tandem repeats in coding DNA include very short nucleotide repeats, moderately sized repeats and very large repeats that can include whole genes. Depending on the degree of sequence homology between the repeats, tandem repeats are liable to a variety of different genetic mechanisms causing sequence exchange between the repeats.

Often such sequence exchanges result in changes in the number of tandem repeats. A reduction in repeat number can often result in a pathogenic deletion, but expansion by sequence duplication can be pathogenic too. Certain chromosomal regions, notably the subtelomeric and pericentromeric regions, harbour large tracts of duplicated DNA and instability of such regions can predispose to disease (Eichler, 1998). Interspersed repeats can also cause pathogenic mutations by a different variety of mechanisms.

Slipped Strand Mispairing of Short Tandem Repeats Predisposes

Insertions and deletions in coding DNA are rare because they usually introduce a translational frameshift. However, occasionally, a series of tandem repeats of a small number of nucleotides occurs by chance in the coding sequence for a polypeptide. Such repeats, like classical microsatellite loci, are comparatively prone to mutation by slipped strand mispairing. As a result, the copy number of tandem repeats is liable to fluctuate, introducing a deletion or an insertion of one or more repeat units. If the mutation occurs in polypeptide-encoding DNA, a resulting deletion will often have a profound effect on gene expression. Frameshifting deletions will normally result in abolition of gene expression. Even if the deletion does not produce a frameshift, deletions of one or more amino acids can still be pathogenic.

Small frameshifting insertions will also be expected to lead to loss of gene expression and often the insertion is a tandem repeat of sequences flanking it. However, nonframeshifting insertions would often not be expected to be pathogenic, unless the insertion occurs in a critically important region, destabilising an essential structure or impeding gene function in some way. Note that large triplet repeat expansions can lead to disease by mechanisms that are not understood at present.

Interspersed Repeats often Predispose to Large Deletions and Duplications

In several cases, the endpoints of deletions are marked by very short direct repeats. For example, the breakpoints in numerous pathogenic deletions in mtDNA occur at perfect or almost perfect short direct repeats. Of these, the most common is a deletion of 4977 bp which has been found in multiple patients with Kearns-Sayre syndrome, an encephalomyopathy characterised by external ophthalmoplegia, ptosis, ataxia and cataract.

The deletion results in elimination of the intervening sequence between two perfect 13 bp repeats and loss of the sequence of one of the repeats. The mitochondrial genome is recombination-deficient and Shoffner *et al.* 1989 have postulated that such deletions arise by a replication slippage mechanism, similar to that occurring at short tandem repeats. Partial duplications of the mitochondrial genome are also distinctive features of certain diseases, notably Kearns-Sayre syndrome. The ends of the duplicated sequences, like those of the common deletions, are often marked by short direct repeats, and the mechanisms of duplication and deletion appear to be closely related.

The Alu Repeat as a Recombination Hotspot

Some large-scale deletions and insertions may be generated by pairing of nonallelic interspersed repeats, followed by breakage and rejoining of chromatid fragments. For example, the *Alu* repeat occurs approximately once every 4 kb and mispairing between such repeats has been suggested to be a frequent cause of deletions and duplications. Some large genes have many internal *Alu* sequences in their introns or untranslated sequences, making them liable to frequent internal deletions and duplications. For example, the 45-kb low density lipoprotein receptor gene has a relatively high density of *Alu* repeats (approximately one every 1.6 kb). A very high frequency of pathogenic deletions in this gene are likely to involve an *Alu* repeat, usually at both endpoints, and occasional pathogenic intragenic duplications also involve *Alu* repeats (Hobbs *et al.*, 1990). Such observations have suggested a general role for *Alu* sequences in promoting recombination and recombination-like events. Initial gene duplications in the evolution of clustered multigene families may often have involved an unequal crossover event between *Alu* repeats or other dispersed repetitive elements. It should be noted, however, that some *Alu*-rich genes do not appear to be loci for frequent *Alu*-mediated recombination.

Physical and Transcript Mapping

A wide variety of physical mapping strategies have been used to analyse the DNA of complex eukaryotic genomes. In the following sections, an arbitrary distinction is made between two classes of physical mapping:

- low resolution physical mapping - the smallest map unit that can be resolved is typically one to several megabases of DNA;
- high resolution physical mapping - the resolution is typically very high, from hundreds of kilobases to a single nucleotide.

Since mammalian DNA has only a very small percentage of coding DNA (~3% in the case of the human genome), a variety of transcript mapping methods have been developed for selectively identifying and studying transcribed sequences, the most important and, arguably, most interesting component of the genome. 1Low resolution physical mapping

Somatic cell Hybrid Panels can Permit Chromosomal Localisation of any Human DNA Sequence

Under certain experimental conditions, cultured cells from different species can be induced to fuse together, thereby generating somatic cell hybrids. In human genetic mapping, hybrid cells are typically constructed by fusing human cells and mouse or hamster cells. The initial fusion products are described as heterokaryons because the cells contain both a human and a rodent nucleus. Eventually, heterokaryons proceed to mitosis, and the two nuclear envelopes dissolve. Thereafter, the human and rodent chromosomes are brought together in a single nucleus. The hybrid cells are unstable. For reasons that remain unknown, most human chromosomes fail to replicate in subsequent rounds of cell division and are lost. This gives rise, eventually, to a variety of more-or-less stable hybrid cell lines, each with the full set of rodent chromosomes plus a few human chromosomes. The loss of the human chromosomes occurs essentially at random, but can be controlled by selection.

The human chromosomes in somatic cell hybrids can conveniently be identified by PCR screening with sets of chromosome-specific primers (Abbott and Povey, 1991). By collecting hybrid cell lines with different human chromosome contents it is possible to generate a hybrid cell panel that can be used to map any human DNA sequence to a specific chromosome. To do this, each of the hybrid cell lines is tested for the presence of the human sequence of interest.

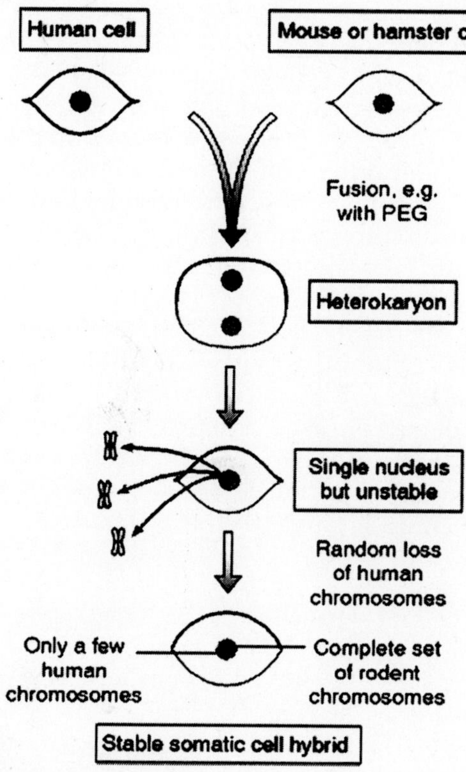

Figure: *Fusion of cells from different species can result in stable somatic cell hybrids.*

A PCR assay can be used with primers specific for that sequence (this is a form of STS content mapping) or the relevant

DNA sequence can be labelled and used as a hybridisation probe. Both of these methods require that the human sequence should be distinguishable from any homologous rodent sequence. Somatic cell hybrids could also be used to establish which chromosome encodes a human enzyme or other gene product by biochemical typing of the hybrids. Whichever human chromosome contains the sequence of interest, it will be present in all the cell lines that are positive for the test and absent from all cell lines that are negative for the test.

Monochromosomal hybrids generated by microcell fusion allow direct and rapid chromosomal localisation of human DNA clones. Traditional somatic cell hybrids generally contain several human chromosomes. Microcell fusion (Fournier and Ruddle, 1977) can be used to generate monochromosomal hybrids that contain only a single human chromosome. The donor human cells are subjected to prolonged mitotic arrest by continued exposure to an inhibitor of mitotic spindle formation, such as colcemid. This causes the chromosome content of a cell to become partitioned into discrete subnuclear packets (micronuclei). Centrifugation in the presence of cytochalasin B (a mitotic spindle inhibitor), results in the formation of microcells. Microcells consist of a single micronucleus with a thin rim of cytoplasm, surrounded by an intact plasma membrane. In much the same way as for normal donor cells, the microcells can be fused with recipient rodent cells to generate hybrids (microcell-mediated chromosome transfer). Some microcell hybrids contain a few donor chromosomes, but the simplest contain a single donor chromosome.

Bibliography

Alberts B, Johnson A: *Molecular Biology of the Cell*, New York: Garland Science, 2022.

Altered Fates: *Gene Therapy and the Retooling of Human Life*, W. W. Norton and Company, New York, 2015.

Banki, L.: *Bioassay of Pesticides in the Laboratory*, Akademiai Kiado, Budapest, 2008.

Bruce, Donald : *Engineering Genesis: The Ethics of Genetic Engineering in Non-Human Animals*, London: Earthscan, 2018.

Cavalli-Sforza, L.L., P. Menozzi: *The History and Geography of Human Genes*. Princeton: Princeton University Press, 2014.

Devi, C.R. Sudharmai : *Analytical Procedures in Soil Science and Agricultural Chemistry*, Agrotech, Delhi, 2017.

Elliott, B N : *Biochemistry and Molecular Biology*, Oxford University Press, Delhi, 2015.

Flores E,: *The Cyanobacteria: Molecular Biology, Genomics and Evolution* (1st ed.). Caister Academic Press, 2023.

Forbes, W. T. M. : *The origin of wings and venational types in insects*, American Midland Naturalist, 1993.

Gay, W.I. : *Methods of Animal Experimentation*, Academic Press, New York, 2005.

Gorelick, F. : *Did insect pollination cause increased seed plant diversity?* Biological Journal of the Linnean Society, 2011.

Halacy, D.S., Jr.: *Genetic Revolution, Shaping Life for Tomorrow,* Harper & Row, Publishers, New York, 1994.

Heller C. :*Life: The Science of Biology*, Massachusetts: Sinauer Associates, 2012.

Jones E: *Genetics: Analysis of Genes and Genomes,* Jones & Bartlett, 2015.

Knudtson, P.: *Genethics: The Clash Between the New Genetics and Human Values*, Harvard University Press, Cambridge, Mass, 2009.

Landa, Norbert: *Ingenious Genes Microexplorers*, Barron's, Hauppauge, New York, 2007.

Lewontin, R.C.: *The Doctrine of DNA: Biology as Ideology*. Harmondsworth: Penguin, 2013.

Lorenzen, S.: *The Phylogenetic Systematics of Freeliving Nematodes*, London, The Ray Society, 2014.

Morris, Jonathan : *Ethics and Biotechnology*, New York: Chelsea House, 2019.

Nisha Khalsa: *Essentials of Biochemistry*, Aavishkar Pub, Delhi, 2008.

Primrose, S. B. : *Molecular Biotechnology,* Blackwell Scientific Publications, Boston, 2011.

Sapp, J.: *Evolution by Association: A History of Symbiosis*. Oxford University Press, UK, 2014.

Singh, Bimla: *Biochemistry and Genomic Revolution*, Vista International Pub, Delhi, 2016.

Swarnim, K. : *A Textbook of Biochemistry and Microbiology*, Surendra Pub, Delhi, 2020.

Tripathi, Nirmal : *Molecular Biotechnology,* Crescent Publishing Corporation, Delhi, 2018.

Valdes, L.A.: *The DNA of God?* Hodder and Stoughton, London, 2008.

Vasudev Bhatnagar: *Cell Science and Technology*, Campus Books, Delhi, 2019.

Wilson, I.: *The Blood and The Shroud*, Weidenfeld & Nicolson, London, 2018.

Winston, R.: *The Future of Genetic Manipulation*, Phoenix, London, 2007.

Wright, L.: *Twins: Genes, Environment and the Mystery of Human Identity*, Weidenfeld & Nicolson, London, 2014.

Index

❑❑❑